Louis Compton Miall

Round the Year : a Series of short nature-Studies

Louis Compton Miall

Round the Year : a Series of short nature-Studies

ISBN/EAN: 9783337024659

Printed in Europe, USA, Canada, Australia, Japan

Cover: Foto ©berggeist007 / pixelio.de

More available books at **www.hansebooks.com**

ROUND THE YEAR

A SERIES OF

SHORT NATURE-STUDIES

BY
PROFESSOR L. C. MIALL, F.R.S.

WITH ILLUSTRATIONS CHIEFLY BY A. R. HAMMOND, F.L.S.

London
MACMILLAN AND CO., Limited
NEW YORK: THE MACMILLAN COMPANY
1898

RICHARD CLAY AND SONS, LIMITED.
LONDON AND BUNGAY.

First Edition, 1896.
Reprinted, 1898

PREFACE

I HAVE given the title of " Round the Year " to a series of sketches suggested by the natural events of the year 1895. The principles of selection have been simple: I have written upon things which happened to interest me at the time, which seemed to admit of popular treatment, and which had not been fully discussed, so far as I knew, in elementary books. Another naturalist would have made a different choice; all the naturalists in the world could not exhaust the subject.

The readers whom I hope to find are observers (especially young observers) of out-of-door nature, teachers of elementary science, and all who care for Live Natural History.

L. C. M.

LEEDS, *June*, 1896.

866905

TABLE OF CONTENTS

	PAGE
"MAN SIEHT NUR WAS MAN WEISS"	1
INSECTS AND PLANTS IN MID-WINTER	3
WHITE OF SELBORNE	8
SNOW-FLAKES	11
BURIED IN THE SNOW	25
BIRDS IN MID-WINTER	26
THE DEPTH TO WHICH THE GROUND FREEZES	28
THE GREAT FROST OF 1895	30
UNDER THE CRAGS	32
PHI AND THETA	38
WHICH ARE THE WETTEST MONTHS?	47
ANIMALS WITH AND WITHOUT COMBS	49
THE MOON	54
SPRING CROCUSES	64
CATKINS	73
THE OIL-BEETLE (MELOE)	89
THE CORN-RIGS OF BEAMSLEY FELL	103

TABLE OF CONTENTS

	PAGE
THE CUCKOO	107
BUDS	121
THE BOTANY OF A RAILWAY-STATION	137
SUMMER TWILIGHT	140
MIDSUMMER BLOOMS	142
HAY-TIME	143
THE HISTORY OF THE CABBAGE WHITE BUTTERFLIES	158
CABBAGES AND TURNIPS	183
DUCKWEED	192
ROUTINE	199
WEEDS	200
MOORLAND PLANTS	208
THE LOVE OF MOUNTAINS	229
THE REVERSED SPIRAL	236
GOSSAMER	240
FLOWER-HAUNTING INSECTS	248
TENNYSON AS A NATURALIST	252
THE STRUCTURE OF A FEATHER	259
THE FALL OF THE LEAF	270
AUTUMN WINDS AND WINTER FLOODS	278
THE SHORTEST DAY OF THE YEAR	288

ROUND THE YEAR

"MAN SIEHT NUR WAS MAN WEISS."

Jan. 4, 1895.—As I light my lamp, and sit down to write, a cold north-east wind is whistling round the house. Thin snow whitens the hills, except where the woods and hedges stand out as black patches and lines. The river Wharfe and the little brooks which flow down from the moors are edged with ice. The sun has just set. To-day the moon completes her first quarter, and is now high in the clear sky. Mars is faintly shining in the south, not far from the moon, and in the east I see Jupiter. In another hour Jupiter will be brilliant indeed.

The earth seems still, and cold, and dead. Yet there are living things hidden everywhere around. This morning my boys found a live caterpillar of the Yellow Underwing, lying helpless on the snow, driven out, perhaps, from its underground retreat by the cold. The experienced collector can find plenty of pupæ, even in the depth of winter. They are hidden away beneath wall-copings, on palisades, beneath loose bark, in moss, or underground. A keen eye is wanted to distinguish them, for their colouring is strongly imitative, and their retreats carefully chosen. Nor will keen eyes suffice unless there is knowledge also. "*Man sieht nur was man weiss,*" says Goethe, and the insect-hunter verifies the saying. The trained naturalist goes about in the winter, and sees living things everywhere. The eager, but uninstructed naturalist can hardly find anything.

Many years ago I got a practical lesson on this subject. I was visiting at a country house in Craven, and the lady of the house showed me her beautiful fernery with some pride. "I am anxious to get the Adder's tongue fern," she said, "but I have hunted for it in vain." I knew that there were some likely meadows at no great distance, and proposed a walk. We went two or three miles, and by groping among the mowing grass soon found ten or twenty plants. The only difficulty was to distinguish the leaf of the Adder's tongue from the rather similar leaves of Hawkweed and Daisy. The plants were packed up, and we walked back. In every field, now that our eyes were opened, we saw the Adder's tongue, and said with some amusement: "If we had only looked as we walked along, we might have saved ourselves the trouble of a long walk." At last we entered the grounds again, and on the lawn, five yards from the front door, there was as much Adder's tongue as could be desired.

Moonwort is another little fern, which is reputed rare. In Yorkshire it grows abundantly on certain stony pastures, often at a considerable height, and would be considered no uncommon plant, if it were only easy to distinguish.

INSECTS AND PLANTS IN MID-WINTER.

Yesterday I walked to Barden Tower to find some Simulium larvæ. These little black creatures, from one-eighth to half an inch long, cluster on leaves of water-cress and brooklime in a clear and rapid stream, which flows down from the moors to the Wharfe. The manœuvres of this larva have been a favourite study of mine. I have watched it clinging to smooth

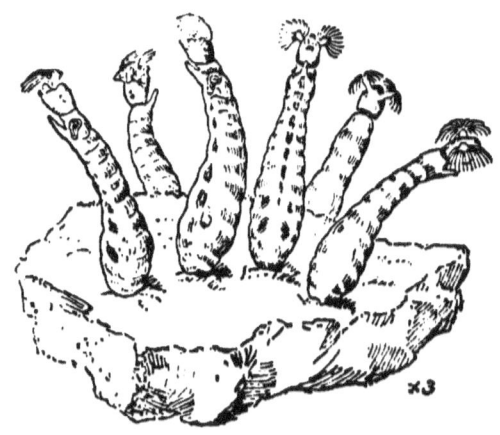

FIG. 1.—Group of larvæ of Simulium attached to a stone.

leaves or stones, in spite of the full force of a mountain current. It keeps its hold by means of a sucker armed with a circle of hooks at the tail-end of its body, or by a somewhat similar sucker just behind the head. If compelled to let go by threatening danger, it disappears from view in a moment, but the attentive observer can by and by see it wavering in the clear torrent, and then slowly travelling back, not by swimming, but by hauling itself in along a thread, one of a number of threads which are stretched from leaf to leaf like those of a Spider, a Geometer, or a Tortrix larva. All this I have described in detail in my *Natural History of Aquatic Insects*. A few days ago a brother-naturalist, a most careful observer of Insects, wrote to express his complete disbelief in the suckers of the Simulium larva. The hooks and threads were there, but no suckers were required or supplied. In some disquiet, for my description was already in type, I set off to procure a few fresh larvæ, and repeat my former observations. It was a clear frosty morning, but the Meteorological Office promised us a gale, so I took my waterproof. The road through Bolton Woods was frozen hard, and made an excellent slide in level places where the rain-water had turned to ice. Before noon the sky was overcast, and a south-west wind blowing, with plenty of warm rain. However I reached the brook, and got as many larvæ as I wanted. They were motionless, though clinging as firmly as ever to their leaves. It was not till they had been half an hour in my warm study that they revived and crept about. Then I picked them up, one by one, with a camel-hair pencil, and put them into a clean saucer full of water. They adhered in a moment, and crept about like Leeches, applying the fore and hind sucker alternately to the smooth porcelain. Time after time I repeated the trial, and it never failed. At last I transferred a larva to a clean slip of glass, and held it under the full stream from a tap. It was not dislodged, and then I was persuaded that the suckers were real, and not imaginary.

Cold seems to have benumbed even the hardy

Simulium larvæ. Insects of all kinds which pass the winter as larva, pupa, or imago are nearly always motionless in very cold weather. Yet not quite always. I have seen (and many other naturalists have seen the same) the great Water-beetle, Dytiscus, swimming about beneath the ice on which I was skating. How do the motionless pupæ, sticking to

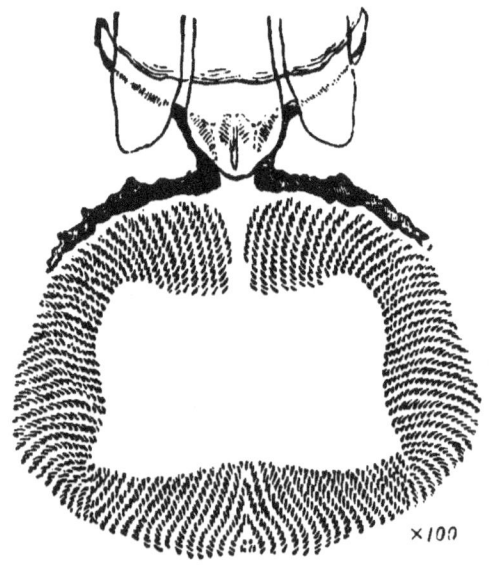

FIG. 2.—Coronet of hooks at tail-end of Simulium larva.

the bark of a tree or to a gate-post, escape being frozen?

Gilbert White, in his account of the great frost of 1776, says that a thaw set in on the 1st of February, "and on the 3rd swarms of little insects were frisking and sporting in a court-yard at South Lambeth, as if they had felt no frost. Why the juices in the small bodies and smaller limbs of such minute beings are not frozen is a matter of curious enquiry." I suspect that his little Insects were Diptera, such as Psychoda and Trichocera hiemalis, which had escaped the extreme cold by sheltering as pupæ in decaying vegetable matter, and only emerged as flies when more genial conditions had returned.

It is surprising how great a severity of cold can be endured by living plants and animals. Siberian Larches endure a mean January temperature of $-45°$ C., falling to a minimum of $-60°$, and rising to a maximum of only $-28°$. Plants have been known to survive after being covered for four years by the advance of a glacier, and abundant vegetation surrounds, and in places overspreads, the great glaciers of Alaska.[1]

Evergreen leaves are probably protected to some extent by unfreezable contents (essential oils, resin, turpentine, benzine, carotin, etc.). The fluids of wood are contained in capillary tubes, and it is well known that under such conditions even pure water will only freeze at a temperature below $0°$ C. In the same way the very minuteness of certain Insects may be a means of safety in severe cold. We know little of the fluids of plants in the depth of winter, but it is probable that they are then both more scanty and more concentrated than at other times.

Animals can be, to all appearance, frozen hard, and yet revive. Ross found in the Arctic regions pupæ of Colias, which were hard and brittle, but afterwards yielded Butterflies. Pierret observed the same thing

[1] On the endurance of cold by plants see Seward's *Fossil Plants as Tests of Climate*, ch. iii.

in the Lime Hawk-moth, Lacordaire in Leucania, Xambeu in the Goat Moth. More than two hundred years ago Lister had noted that caterpillars and pupæ, though frozen till they became brittle and tinkled against glass, were capable of reviving completely. It has even been found possible to freeze a Frog to a rigid body without destroying life, but the trial rarely succeeds. In all these cases it is probable that a part only of the contained water turns to ice. Complete freezing would remove water from the albumens and other organic compounds of the body, and would be certainly fatal. An extremely low temperature would be necessary. Ten hours' exposure to $-16°$ C. was required to solidify the contents of a Fowl's egg, which were even then, in all probability, only partially frozen.[1] During the freezing of many solutions and mixtures partial solidification sets in first, and the fluid residuum becomes more and more difficult to freeze, as solidification proceeds.

Even when we are aware of the difficulty of freezing an animal completely, we cannot but wonder that Mammals, Birds, Insects, and many other living creatures should survive the winter cold of the far north. I would not undertake to explain how animals which are unable to replenish their store of energy can endure for months together a temperature at which mercury freezes.

The thawing of frozen tissues is in most cases even more dangerous than the freezing itself. Diffusion-currents break up the microscopic structures, and change the composition of the fluids. In the same

[1] Kochs, *Biol. Centralblatt*, 1892, 1895.

way the mere immersion of pieces of living tissue in pure water quickly renders them unfit for microscopic investigation. The quicker the process of thawing the greater the risk. Hence the well-known rule of treating frost-bites by rubbing with snow in the open air has theory as well as experience to support it. The noon-day sun of winter is more deadly to plants and certain peculiarly exposed animals than the midnight frost itself.

WHITE OF SELBORNE.

Gilbert White's name brings up the most delightful recollections. I can, after forty or four hundred readings, take up the *Natural History of Selborne* again, and brighten with it that last hour of the day when work is put aside, and the house is still. What is it which gives this unfailing charm to the memoranda of the quiet old curate of Selborne? First of all, he had a considerable knowledge of his subject, which is a condition not to be dispensed with. Then he was a keen observer and a diligent recorder. How much he rescued from forgetfulness by that habit of noting things down at the time! "Half a word fixed upon or near the spot is worth a cart-load of recollection," says the poet Gray, who was naturalist and antiquary as well as poet. White loves circumstance, and one is, at first sight, inclined to think that he is interesting merely because he gives you all the particulars. Try your own hand at writing about nature, and see whether you get a lively narrative by setting down all the facts, great and small! Voltaire says :—" le secret d'ennuyer est celui de tout dire." White's method is to select and to select carefully the particulars which have human interest; all the rest he leaves out. He displays the skill of the old-fashioned letter-writer, and selects from the particulars of which his memory is full as carefully as Madame de Sévigné. White never forgets that his birds and insects are, or lately were *alive*. Too many naturalists write about them as they might write of Skiddaw or Stonehenge, being, it would seem, chiefly solicitous to note where they are to be found. But White thinks of their hardships and expedients. His moderation and good sense are shown by his keeping well within his own range. Others might, like Buffon, develop their theories of the earth in magnificent rhetoric, but White is content to "stoop to what he understands." Few naturalists of the last century require so little correction or explanation in the present day.

The frosty weather has sent me once more to read White's account of the snow-storms and frosts of 1768, 1776 and 1784. How different is he from the mechanical narrator, to whom all facts are equally interesting! White *thinks* about everything that he notes down. Observe his reflections upon the effect of intermittent cold upon trees, and shrubs, and bees; upon the endurance of cold by small Insects; upon the occurrence of great cold on low ground when it is warmer at places a few hundred feet higher. Notice too, the practical turn of his mind. He bids the planter shake off the snow daily, so as to lessen the damage due to repeated melting and freezing of the snow upon the shrubs. He notes the shrubs which suffered most from frost, in order that his friends may learn to plant only such as can stand severe cold. This is of a piece with his constant interest in household matters, the making of rushlights, the small, long, shining fly which lays its eggs in bacon, the holes gnawed by crickets in stockings and aprons hung to the fire, and the like.

White foretold one of the chief applications of Zoology to the practical affairs of mankind in the following passage :—" A full history of noxious insects hurtful in the field, suggesting all the known and likely means of destroying them, would be a most useful and important work. A knowledge of the properties, economy, propagation, and in short of the life and conversation of these animals, is a necessary step to lead us to some method of preventing their depredations."

A little elementary Physics, so cheap nowadays, would have greatly mended some of White's explanations. He thinks that thaws often originate under ground, from warm vapours that arise. He remarks truly enough that " when a thermometer hangs abroad in a frosty night, the intervention of a cloud shall immediately raise the mercury ten degrees ; and a clear sky shall again compel it to descend to its former gauge." But this leads him to conclude that "cold often seems to descend from above." Nor could he interpret his own observation of unusual cold in low-lying and sheltered spots. It is easy now to point out that in perfectly still weather the air which is chilled, and therefore of greater density, will collect in hollows. Promising boys in an elementary school are taught many things which the observant and well-read Gilbert White never came to know.

Would that we had a constant succession of naturalists of White's sort! Natural History is being choked by unassimilated facts, mechanically compiled by men who have apparently ceased to think about Nature Hence a profuse and growing literature of the most melancholy description, dry, marrowless, useless. We record and record till our catalogues grow too voluminous for storage, and too stodgy for the toughest appetite. Why do we go on printing this stuff? Because a considerable section of the public believes in Natural History, and is willing to pay for much that it never reads. When the purchaser is not a reader, the quality of the writing may sink to any level whatever.

SNOW-FLAKES.

Jan. 6.—Snow is falling thick this morning. I have been out of doors to look at the snow-flakes. All that is required is a plate to catch the snow, and a pocket-lens. The plate must not be above the freezing-point. A sheet of coloured paper often does as well or even better. This morning the snow-crystals were not first-rate. They were large and irregular, several cohering together and blurring one another's outlines. This is usually the case when the air is close to freezing-point and somewhat moist. If my first inspection had been quite satisfactory, I should have brought out a microscope, and allowing time for cooling, should have examined the crystals more carefully, as I have done many times before.

In the northern countries of Europe at least the crystalline form of snow must have been observed very long ago, but nothing was said about it in books till the 12th century, when Albertus Magnus remarked that snow-flakes had the form of a star. Olaus

FIG. 3.—Snow-crystals, photographed by Dr. R. Neuhauss. From Prof. G. Hellmann's *Schneekrystalle.*

Magnus, in his *History of the Northern Nations* (1555) figures snow-flakes, but so wretchedly that he (or his engraver) had hardly recognised that they are always angular, much less that the angles are constant. Kepler in 1611 noted that snow-flakes are six-rayed,

SNOW-FLAKES

and asked why. *Cur autem sexangula?* His most hopeful suggestion was that chemists should find out whether snow contained salts, and if so, what they were. Descartes, to whom Meteorology and Optics owe the first good explanation of the rainbow, figures rather conventionally, several kinds of snow-stars, which he observed at Amsterdam in February 1635. The six-rayed star, branched and unbranched, the rosette, the six-rayed star with intermediate rays, and the pair of lamellar crystals joined by a prismatic rod are all shown, and if we make some allowance for the small size of the figures and the rudeness of the engraving, we may say that nothing is shown which does not occur in nature. Robert Hooke in his *Micrographia* (1665) gave many fairly good figures, and first noticed that in branched snow-stars " the branchings from each side of the stems were parallel to the next stem on that side," an immediate consequence from the fact that the secondary branches, in this case of 60°, make the same angle with the primary axes, as these make with each other. Rosetti of Leghorn in 1681 first observed the extremely minute cavities ("capillary cavities" or "air-spaces" of modern writers) to be found in snow-flakes. Scoresby in 1820 published an elaborate account of the form of the snow-flakes of the Arctic regions, figuring 96 forms, and classifying them under five principal heads. The figures were completed symmetrically, and the dimensions given. Glaisher in 1855 published 150 figures of snow-crystals, all completed symmetrically. Scoresby's and Glaisher's figures have been copied in many common text-books. The next step was to photograph snow-crystals, and this has been done with great success by Neuhauss and Nordenskiöld.[1] It is obvious that photographs record many particulars which cannot be accurately recorded by drawings, made from objects so fugitive as snow-flakes. They also preserve many departures from symmetry which have been neglected in the drawings made by hand.

The large flakes, with a diameter of half or three-quarters of an inch, are not themselves snow-crystals, but aggregations of such, sometimes very loosely attached. A diameter of a fifth of an inch is very large for a snow-crystal, and the average diameter of snow-stars, the commonest form, is only about one-tenth of an inch (2.35 mm.) Other snow-crystals are yet smaller. Keen sight or the help of a lens is therefore necessary to make out the exact shape of a snow-crystal, and all the finer details require the microscope.

When we examine a snow-crystal carefully, we soon learn one fact respecting it, viz., that it is six-rayed. The crystal forms along three lines or axes lying in one plane, which cross each other at equal angles. Six lines proceeding from a common point at equal angles will, of course, be 60° apart. There is a fourth axis to the snow-crystal, which we do not recognise at first. This represents the thickness of the crystal, and takes a direction at right angles to the plane in which the other (*lateral*) axes lie, passing through

[1] Neuhauss's figures are reproduced in Hellmann's *Schnee-krystalle*, a useful and interesting book, which has furnished the materials for this historical sketch. A number of Nordenskiöld's figures are reproduced in facsimile in *Nature*, October 19, 1893.

their place of intersection. The fourth axis is usually so short that the crystal has hardly an appreciable thickness; but it may be long. Some snow-crystals take the form of double stars separated by a prismatic rod longer than the diameter of the stars. Here the prismatic rod lies along the fourth axis, which is often called the *principal axis*.

If we mark points on the three lateral axes, equidistant from their intersection, and join these one to another by straight lines, we shall get a regular hexagon, whose angles are, of course, $\dfrac{8 \times 90°}{6} = 120°$ (Fig. 4.)

Hexagonal plates often enter into snow-crystals, forming the centre, or tipping the rays; some crystals are nothing but hexagonal plates.

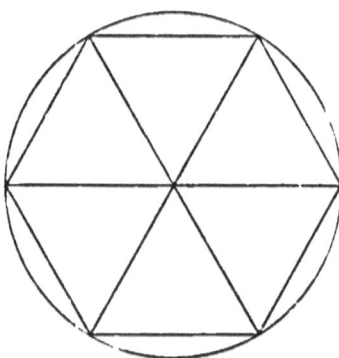

FIG. 4.—A regular hexagon in a circle, with three axes crossing at angles of 60°.

Solid water in all its forms is essentially similar to the snow-crystal. During a thaw six-sided prisms sometimes stand out from the surface of the melting ice. The spicules which shoot across the surface of freezing water make angles of 60°. Hoar-frost, when studied with a lens, is seen to be built up of six-sided crystals.

If we take any two adjacent triangles in Fig. 4 and join them, they will form a four-sided rhomb. Three such rhombs make up the hexagon. Rhombohedrons are often seen to form the side branches of a principal ray, and we could easily build up certain forms of snow-stars entirely out of rhombs. (Fig. 5.) I do not, however, know that snow-crystals are actually so formed throughout, though some show lines of union which point to such an arrangement. I have seen crystals with the centre, others with the rays thus divided. It is uncommon to find the centre built up apparently of rhombs. Far more frequently it consists of a tabular hexagon, often with long rays

FIG. 5 —Snow-star, subdivided into rhombs.

proceeding from its angles, and with ribs or raised lines taking the same direction.

Snow-crystals are very seldom quite regular. Perfect regularity would mean that the air was still, of uniform temperature, and uniformly supplied with moisture. There may, however, be a near approach to such conditions, as the great regularity of some few crystals shows. The fall of crystals through great distances tends to impair their form. They are apt to stick together and to form flakes, or to gather moisture on one side more than on another. When the air is quite still and very cold, a thin mist near the ground sometimes turns to snow. This is the "diamond dust snow" of the Arctic regions, and it is believed to consist of very small and unusually perfect crystals.

We cannot follow by the eye, even when aided by the microscope, all the details of the growth of a snow-crystal. But we can draw or photograph, and try to interpret what we have seen.

The finest particles of liquid water, condensed from water-vapour, attach themselves to solid bodies, as if attracted by them. They seem to be easily attracted through small distances by spicules of ice, and in solidifying they commonly place themselves regularly with respect to particles which have previously crystallised.

Suppose that we have to begin with a single needle of ice, and that the moisture suffices to form more needles. They will shoot out from the sides of the first at angles of 60°, forming a six-rayed star. The angles are often filled up by the shooting across of fresh needles, which make the same angle of 60° with the primary rays, and thus a regular hexagon is formed. It is not uncommon to find among snow-crystals very perfect and simple hexagons. If they grow, they will probably send out rays from their corners. Why from the corners rather than from their flat sides? Because the corners project farther into the field, and are more exposed to the contact

C

of floating particles. Let us suppose that each corner has gathered its quota of particles, and that these have arranged themselves regularly along rays pointing towards the centre of the hexagon, and making angles of 60° with one another. When the rays have pushed out a good way from the centre they will attract very feebly some of the floating particles which lie between them. Secondary rays will then dart out into the midst of the particles from the primary rays, making the same angles as before. By a repetition of the process we may get a star of any degree of complexity, and if the particles of moisture are uniformly distributed, its symmetry will be perfect. But if there is more moisture or greater cold in one part than in another, perfect symmetry will be lost. The branching rays will shoot, but nearly always at constant angles, into the patches of moisture and avoid dry places, just as the growing branches of a tree push into the sunshine and avoid the shade. Sometimes, however, the original direction of the rays is not quite accurately preserved. Crystal joins to crystal, not in straight lines, but with slight deviations. This is not the case, so far as I know, with snow-crystals, but it is common in hoar-frost and on the frosted pane. The neighbourhood of a solid body may possibly set up these disturbances.

When we have got a central hexagon with rays attached to its angles—a common, but by no means inevitable form, there is often a change in the angle of attachment of the secondary rays. One of the sides of the hexagon passes from one primary ray to

SNOW-FLAKES

another in such a direction as to give an angle of 60° on its inner side (nearest to the centre) and an angle of 120° on its outer side. But at a distance from the centre of the crystal, we more commonly find the angle of 120° inside, and the angle of 60° outside. The angles which the more distant secondary rays make are in such cases *supplementary* to those of the more central secondary rays. But the angle of 60° is still there, though its place may be changed.

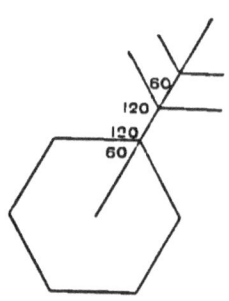

FIG. 6.—Angles of 120° and 60°. The secondary axes are turned inwards near the point of intersection, and outwards at a greater distance.

We may say that near the point of intersection of the primaries the secondary rays are turned as nearly towards the centre as their angle of 60° allows, while at a greater distance they are turned away from the centre. Why is this? The first arrangement is the most compact, the other the least compact possible. Near the centre there is at first an excess of moisture, and here the rays will be crowded, often forming a solid mass; all turn *towards* the centre. As the rays shoot further and further out the moisture becomes less plentiful; it was presumably less plentiful from the first, and the growth of the crystal has further diminished it. Now the new rays will take that arrangement which spaces them most widely; they will all turn *from* the centre.

It is a good plan to draw all snow-crystals that are observed. A pair of compasses and a ruler are wanted. If you wish to draw a regular hexagon, notice that the radius of the circle which bounds the hexagon is exactly the length of one of the sides. It is convenient to have ready a card cut to the figure of a rhomb, with angles of 60° and 120°. These can be got from the hexagon. Lines drawn from the angles to the centre meet at 60°, and each angle of the hexagon is 120°. As a rule, half of one ray of the crystal is enough to show the crystalline form, and it is generally best to draw no more.

So far we have neglected the thickness of the crystals, and have treated them as flat. But snow-crystals are of three dimensions, and the third dimension is often too large to be neglected. We saw that the angles of the flat figure projected farther from the centre than the rest, and generally gathered more floating particles to themselves. It is the same with the edges and solid angles of the crystal of three dimensions. Suppose a great number of small spheres to cohere into a crystalline form, which for the sake of simplicity we will suppose to be cubical On one of the flat faces each particle will be half immersed and half exposed. The particles along an edge will be one-quarter immersed and three-quarters exposed. The particle at a solid angle will be one-eighth immersed and seven-eighths exposed. The greater the exposure the greater the possibility of attracting floating particles, and this helps us to understand how edges grow faster than flat faces, and solid angles faster than edges.[1] But exceptions to the rule are not uncommon, especially in very small crystals.

[1] Sollas, *Nature*, Dec. 29, 1892.

Hoar-frost often gives us good examples of the tendency of crystals to grow out from their points and edges. When the air is still, very cold, and laden with moisture, a white crystalline growth forms on trees, bushes and grass. If we examine it with a

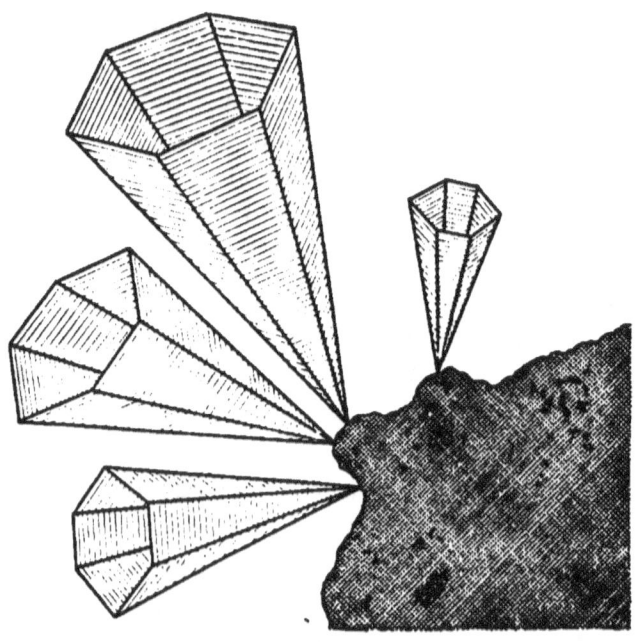

FIG. 7.—Hollow hexagonal "hopper-crystals" of ice, from Surtshellir Cavern, Iceland. From Grossmann and Lomas, *Nature*, Oct. 18, 1894.

lens we shall see many needles, growing at the point. We shall also find hexagonal "hopper-crystals," that is hollow pyramids, attached by the apex and with the cavity turned towards the sky. These are due to more rapid growth of the edges of the crystal. Air-currents and other disturbing causes affect the regularity of the crystalline masses, and the freaks of hoar-frost are many.

Many snow-crystals exhibit under the microscope

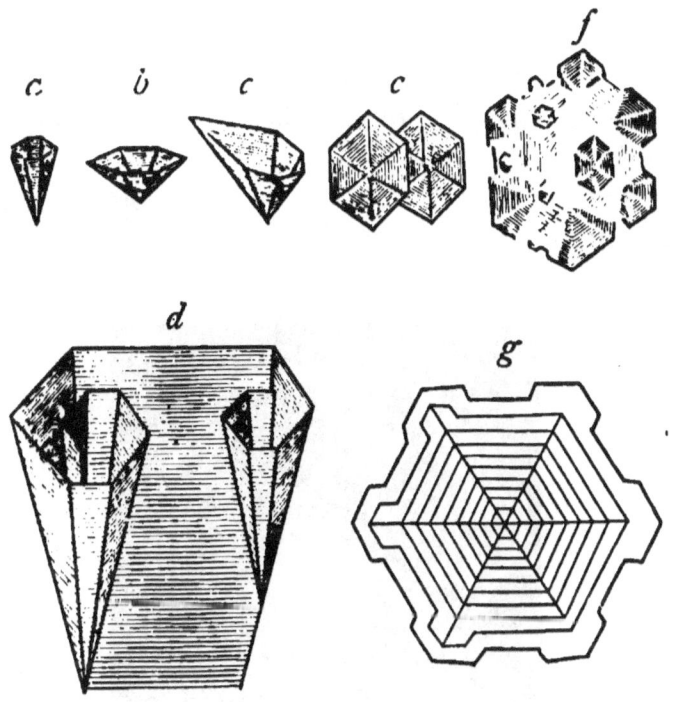

FIG. 8.—Natural hoar-frost, Christmas, 1892 (*a*, *f*, natural size, the rest enlarged). From Grossmann and Lomas, *Nature*, Oct. 18, 1894.

cavities containing air, or water, or both. These cavities are sometimes of very regular form, spheres, or elongated tubes pointed at each end. In other cases three sides of the hexagon exhibit one pattern (*a*. Fig. 9) and alternate with three sides of different pattern (*b*.). The similar crystals of hoar-frost have never been found to contain such cavities.

SNOW-FLAKES

The star-like crystals are believed to form only in great cold, as in the upper regions of the air. Tabular crystals, or flat hexagonal plates, form at higher temperatures. They are common in the lower layers of fallen snow, and appear to the eye as minute lustrous scales. The crystalline form must have undergone change, for tabular crystals rarely fall from the sky.

A sufficiently low temperature will produce snow direct from water-vapour. The story of the Peters-

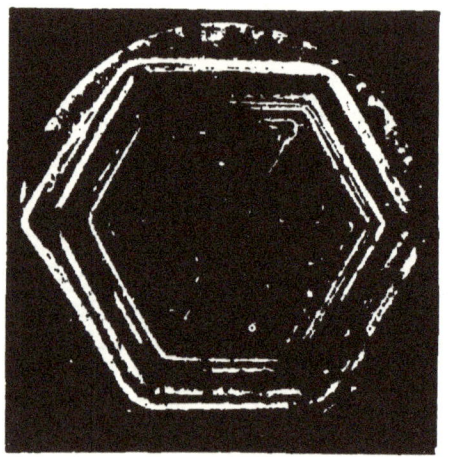

Fig. 9 — Hexagonal snow-crystal, with star-shaped cavity. The alternate triangular fields (*a*, *b*) differ in pattern. Photographed from nature by Nordenskiöld. *Nature*, Oct. 19. 1893.

burg ball-room is well known. The hall was crowded with people and ladies were fainting from heat, when some one opened a window, and the cold air rushing in caused snow to fall. Maupertuis observed in Lapland that the mere opening of a door caused big snow-flakes to fall in the hut. Cold readily forms snow in the vacuum of a water-barometer. Muncke took an exhausted glass bulb and set it, during frost, in the open window of a room, which though not heated, was warmer than the outer air. A little ice had been allowed to form on one side of the interior of the bulb, and this was turned towards the room. Before long the invisible water-particles passed into vapour, and crossing the bulb formed a loose mass of perfectly formed crystals on the opposite side.

Snow, as it lies on the ground, contains much air. Its density is only one-tenth or one-twelfth that of water. Nordenskiöld remarks that even below freezing point snow may contain so much water as to drip.

In ancient and mediæval times it was believed that long-continued and severe cold could squeeze all the liquidity out of water, and that the permanently solid rock-crystal was formed upon the Alps in this way. Rock-crystal is now known to be crystallised silica, which nearly always takes the form of six-sided prisms or pyramids.

Why is snow white? Water, ice, glass and other transparent bodies are not white. But water is white when broken into spray or foam. Ice and glass becomes white when pounded or filled with small air-bubbles. Leeuwenhoeck, two hundred years ago, showed that milk owes its whiteness to minute globules of oil suspended in a watery fluid. A body is white when it reflects much white light. Transparent bodies reflect some light from their surfaces, but allow a great deal to pass through. The more they are broken up, the more numerous do the reflecting surfaces become, and the smaller the quantity of light which is able to pass through.

It will be seen that some progress has been made since Kepler's time in the scientific study of snow, but we are still quite unable to answer that question of his, *Cur autem sexangula?* Why are snow-stars six-pointed?

BURIED IN THE SNOW.

The large quantity of air entangled in loose snow helps us to understand how sheep and even human beings can survive long burial in snow-drifts. Samuel Bowditch, an old writer in the *Philosophical Transactions*, tells the following story :—Joanna Crippen, of Chardstock, in Dorsetshire, a spinner of worsted, went home on the 24th of January (1712?) when it was snowing hard. She lost one of her shoes, and her clothes, which were very poor, were torn by the brambles. At last she lay down under a hedge, it being then about six o'clock on Monday evening. She was not discovered till the following Sunday afternoon, when a party of searchers found her buried four feet deep. A man thrust a pole into the heap when she cried out and begged him not to push her so hard. When dug out she had no shoes or stockings on. Her clothes were very scanty. Her shoulders were covered by an old whittle[1] in which she had gnawed a large hole. She had drunk the snow which melted on her body to quench her thirst.

[1] A *whittle* was a piece of *white* (undyed) cloth, or blanket.

One of her great toes was mortified (frost-bitten?), but she soon recovered, and at the time of writing was described as 'very hearty.'[1] It is well known that the inmates of dwellings buried deep in snow by avalanches have survived for several weeks in more than one case.

BIRDS IN MID-WINTER.

The snow has driven the Red-breasts towards the habitations of man. Some of them are quite fearless, and hop about within a few feet of a window at which faces appear. Others are shy, and keep aloof except when pressed by hunger. They are solitary birds, and are never seen in flocks, rarely two together, except when mating. The old proverb, "One bush does not lodge two Red-breasts," is very fairly true. Naturalists have studied with care the limited migrations of the Red-breasts. They travel south in autumn and return in spring, but in England or Central Europe, Red-breasts are to be found throughout the year. Some go and others take their place; some arrive and others depart.

A few days ago I heard a Red-breast singing lustily out of a leafless tree. There is no time of year when they are silent, but in spring their sweet and varied, though not powerful song, is drowned in the chorus of newly-returned songsters. Even in August, which White calls "the most mute month the spring, summer, and autumn through," the Red-breast "tunes his merry note." Black-birds, Sky-larks, and

[1] *Philosophical Transactions*, No. 337, p. 265 (1713).

Song-thrushes occasionally sing in the depth of winter, and the less melodious noise of the Missel-thrush, the Wren, the Starling, the Hedge Sparrow, the Chaffinch, the Yellowhammer, the Corn Bunting and the Tits may be heard both early and late, in some cases all round the year. I believe that the Missel-thrush never sings except in winter or early spring.

We have few birds as yet about our new-built house, but we shall have more before long. The boys have thrown out corn, and bread, and sand for the starving birds Sand is as necessary as food in snowy weather, for a bird with an empty gizzard cannot digest its food.[1] When our shrubs and trees have grown there will be better shelter, and that will greatly increase the number of our visitors

Birds endure great hardships when the ground is covered with snow for many days together, but I fancy that they care little for mere cold. Such as are fond of bathing will bathe in an ice-cold spring on a frosty morning, and you will rarely find a bird of any kind seeking shelter from a cutting north-east wind. Rain is a different thing. Many birds do not like to get their plumage wet. There is sometimes talk of birds perishing from cold, but it will generally be found by close inquiry into the circumstances that they were short of food and water when they succumbed.

Grouse were plentiful after the hard winter of 1895, but in January, 1886, they suffered greatly. Repeated falls of snow and alternations of frost and thaw

[1] The sand should be coarse and sharp.

covered the ground with a thick and solid frozen mass. The birds strayed in great numbers into the valley of the Wharfe, and were found in the cultivated fields along the river. Many were said to be injured by dashing against the telegraph wires.[1]

THE DEPTH TO WHICH THE GROUND FREEZES.

Jan. 9.—A question came up to-day which I was unable to answer off-hand. This is our first winter in a new house, and the domestic management wants to know whether our water-main, which lies about 2 ft. 6 in. below the surface of the ground, is liable to freeze. I have never had occasion to consider this question, and was, at first, unable to give a clear answer. By and by it occurred to me that the insect larvæ, which winter in the ground, are often found less than a foot deep. It seems probable, therefore, that frost does not usually penetrate to the depth of one foot.

Jan. 11.—Mr. G. J. Symons, F.R.S., who has made long and careful observations of underground temperature, gives me in a letter received this morning some interesting particulars. " The earth," he says, " at one foot below its surface very rarely reaches 32° F. Here (Camden Square, London) it fell to 32·2° on Jan. 6 and 7, 1893, and to 32° on Jan. 29, 31, and Feb. 1 to 5, 1880. It did not fall to 32° *here* in 1875, but it just reached it at the gardens of the Royal Botanic Society, Regent's Park, on Jan. 1, 2, 3 and 4,

[1] *Zoologist*, March, 1886.

DEPTH TO WHICH THE GROUND FREEZES 29

1875. Mains for water supply are usually put 2 ft. below the surface to avoid all risk, and I remember, but cannot give details, that in a small Hertfordshire town an engineer did the economical and put his pipes at one foot. A big frost came, and every main was cracked!'

It is a comfort to know that our water-main is tolerably safe. As to underground larvæ, Mr. Symons adds :—" I should *think* that larvæ in walls, etc., generally get below 32°, but that those that get down 6 in. underground very rarely do so—perhaps one year in five."

So I wrote on Jan. 11. The next few weeks were instructive. In the first place the underground temperatures of the spring of 1895 greatly extended the maximum depth at which a freezing temperature was ascertained. The thermometers showed that

Frost penetrated to	1 ft.	at 11 stations.
,,	,,	1 ft. 6 in.	3 ,,
,,	,,	2 ft.	1 station.
,,	,,	2 ft. 6 in.	no station.

While this was the story told by the thermometers, the water-mains indicated an even greater penetration of the frost. At Maidenhead, Hatfield and Shrewsbury the mains froze at 2 ft. 6 in., and at Musselburgh at over 3 ft. The damage done all over the country was very great. In Liverpool 27,000 houses were without water at one time. In Sheffield nearly 170,000 persons were without proper water-supply. Selkirk reported that so many bursts had occurred that no attempt would be made to repair, but new mains would be laid throughout. From Devonshire to Inverness the same tale of trouble was heard.[1]

Mr. Symons explains the discrepancy between the thermometers and the mains in this way. The frost never really penetrated the soil to pipes at such depths as 2 ft. 6 in., but water cooled almost to freezing-point in uncovered reservoirs was steadily delivered into the pipes, and chilled the surrounding soil. This of itself would not have frozen the water in the mains, but further loss of heat was experienced by conduction along the shallow service-pipes and the pipes supplying the hydrants in the streets. These effected a perfect metallic communication between the mains and the surface of the ground at a time when the air for a long time together was little above zero.

The service-pipes are led off from the top of the mains. If they could be led off from the sides the risk of freezing would be appreciably reduced, but the cost and labour of connection would be materially increased.

Our mains never froze at all, and as the frost of 1895 is believed to be the hardest for eighty years, we shall face future frosts with a light heart.

The Great Frost of 1895.

The frost lasted from Dec. 29 to Mar. 5 (nine and a half weeks), but was broken by a mild interval

[1] Symons's *Monthly Meteorological Magazine* for April, May, and June, 1895, contains much information respecting the frost and its effects, of which I have made use here.

(Jan. 14-21). For seventy consecutive days (Dec. 26 to Mar. 5) the thermometer fell below freezing-point during some part of the twenty-four hours. The mean temperature for the whole period was 27·5°, and the daily minimum averaged 22°. The greatest extremity of cold observed in England was $-11°$ at Buxton. Northerly winds prevailed during most of the time, an area of high pressure being pretty constant in the north, and an area of low pressure in the south of Europe. The ice on ponds attained a thickness of over ten inches. On the whole the frost was considered the most severe since that of 1814.[1]

On our commons and moors the Furze was much injured, and the following summer many dead or partly dead bushes were seen. In Yorkshire Furze is near its northern limit; it gets into the North of Scotland, but its frequency diminishes greatly. On the Continent it does not extend north of Denmark.

Fruit-eating Birds, especially the Black-bird, Thrush, and Linnet, were so reduced in numbers that in the summer of 1895 little damage was done to fruit, and nets were laid aside which had previously been absolutely necessary to protect the fruit-crops. Mountain Ash berries, in places where the trees are ordinarily cleared by the birds every autumn, remained unmolested in singular profusion.

Even in the sea the effects of the long and severe cold were felt. Oysters and Mussels suffered much from hard frosts at low tides. Whelks and Scallops were killed in great numbers. Fishes and other marine animals were thrown up on the shore, dead or

[1] Bayard and Marriott, Roy. Met. Soc., 1895.

helpless. In the following season it was observed by the French fishermen that nearly all the large Shrimps had been killed, and that only young and small ones could be fished. The mortality extended to considerable depths (15-25 metres). It was observed that animals which are usually found only in water of fair depth came close to shore during the frost.[1]

There is a general impression that injurious insects are kept in check by severe winters, but I know of no direct and extensive evidence in support of the view. The hard winter of 1894-5 was followed by a season in which insects, whether injurious or not, showed no diminution of numbers.[2]

Under the Crags.

I have so much dry information to pour out that I will venture to waste a page or two upon my own surroundings. The chapter will be short, and will contain little useful matter.

We live on a steep slope which runs down from Rumbalds Moor to the Wharfe. Here the river flows from west to east, and we are on the southern bank, facing due north. The skyline behind the house is rugged with cliffs and fallen blocks of sandstone, among which are the locally famous Cow and Calf. Below these is a fringe of moor, overgrown with heather, crowberry and moss. Then comes a narrow strip of pasture, on the low side of which is our garden fence. Below us is a rolling mass of grassy

[1] Fauvel *Comptes Rendus*, CXXI. pp. 427-429 (1895).
[2] Miss Ormerod's *Report on Injurious Insects for* 1895.

hillocks, which rest upon a gentle slope. Across the river is a great hillside, six miles long as seen from our windows, which culminates to the west in Beamsley Beacon ; the higher parts are heathery, the lower slopes covered with wood and pasture.

The two sides of the valley might be called Security and Desolation. The opposite slope is stable, and will never move unless all Yorkshire is shaken. Our side is wild and rugged, because of great landslips. We have the better prospects and the more picturesque rocks, but the other side gets the best of the sunshine. In December the sun never shines upon our house for two full hours in the day.

From the Cow and Calf to a line well below our house the whole hillside has slipped. The form of the ground tells the tale. The great sandstone cliffs have been cracked through, and the insufficiently supported edge has fallen in wild ruin, pushing before it great mounds of shale and clay. Near the line of fracture great fissures run through the sandstone, as if more would fall some day. The Calf is on the slipped ground, the Cow is part of the cliff which has stood firm. A quarter of a mile to the west of us the fallen rocks are piled into a long and steep ridge. Between them and the cliffs from which they have broken away lies a considerable hollow, called the Rocky Valley. Eastwards the slip increases in volume, and covers almost all the hillside as seen from the river.

What brought down this great sheet of rocks and earth, which measures more than a square mile in extent? One usual condition of a landslip is conspicuous here—a steep slope of comparatively soft rock (shale) surmounted by a thick bed of hard rock (sandstone). If the strata were to dip ever so little outwards, towards the river, the fall would be hastened. I cannot say that there is any marked outward dip here, for the sandstone is irregularly bedded and shaken, so that no good observation can be made, but there is a marked dip towards the head of the valley. Did the river ever wash the base of the slope and so undermine the cliffs? I think that this was not the cause. Below the hummocks of slipped shale and clay comes a gentle slope, which has never been disturbed or cut into by the river. It is worth notice that the whole landslip is full of water. Springs break out all over its surface, and the rough pastures are never dry.

No history, no tradition of the great slip is preserved; it may have taken place, for all that we know, many thousand years ago, before England was peopled at all. The great lapse of time during which the surface has remained unchanged is our chief reason for living tranquilly on the scene of so great ruin.

Near to the Cow and Calf are sandstone quarries, chiefly interesting for the planing and scratching of ice upon the bared surfaces of the rock. Some of these ice-planes, as well as the big "day-stones" which lie around, bear the rude sculptures known as cup-and-ring marks. When and why they were made no one knows, though antiquarian conjecture is profuse. Flint chips and stone tools are occasionally picked up, while circles and cairns are plentiful on the moor, the relics of tribes whose name has perished.

Fig. 10.—The Cow and Calf, Ilkley.

The allurements of the spot are pure and bracing air, wide prospects, and constant incitements to exercise. The naturalist finds a profusion of flowers and streams swarming with aquatic insects. Mr. McLachlan [1] witnessed an extraordinary spectacle in our valley in the end of September, 1873. All the way from Ilkley to Bolton Abbey (about seven miles) the valley swarmed with a rather uncommon Caddis fly (*Halesus auricollis*). When a branch was shaken the air became alive with the flies, and they covered the grass. I think I have never found any stream quite so full of aquatic insects as Beamsley Beck, which empties itself into the Wharfe close to Bolton Bridge.

We have hill and valley; moor, woodland, meadow and pasture; endless streams and fountains. Few places in England offer more variety, and the naturalist's great danger is that of distraction. Every day there is some new thing to note, and he is apt to become hasty and desultory.

On this bleak slope, which faces due north, and is open to every wind which has north in it, we can grow few plants to advantage. When I was putting in trees and shrubs I ventured to aspire to Apple trees. "Apples?" said the nurseryman. "Yes, you might get blossoms in a good season, and of course there would be the leaves to look at." I climbed down, and contented myself with Gooseberries and Black Currants. Our Cabbages and Lettuces are a success, and the rockery is gay with Alpines, but we attempt nothing that fears the wind.

[1] *Entom. Month. Mag.*, Vol. X. p. 140 (1873).

How is it that the Sycamore endures our wintry gales so bravely? There are many well-grown Sycamores around, both young and old, with shapely heads, and no swerving to one side. Elms, oaks and ashes all bend away from the west. If we were to judge from the Sycamores alone, we should be inclined to say that the wind never raged along this hill-side. The Sycamore is a true Alpine tree, ascending to over 5,000 feet in Switzerland and other mountainous countries.

The place is at its best on a fine summer evening. The sun then sets behind Barden Moor, and his rays are reflected to us from a sickle-shaped bend of the river. The low, square church-tower is bathed in ruddy light. On such an evening it is delightful to sit upon our terrace and watch the colours on the hills change and fade, till the long ridges of moorland stand out black against the still luminous sky.

Some faint historic flavour clings to Denton, a little village across the river, for it was one of the homes of the Fairfaxes. Sir Thomas Fairfax chanced to be born here in 1612, his father then usually dwelling in a house which still stands in the adjoining valley of the Washburn. The old hall came into the hands of a family of Leeds clothiers, who rebuilt it in 1760. The present Denton Hall is handsome and well-placed, but has no historical associations. Middleton Hall is a manor-house of a kind not uncommon in Yorkshire. Though not remarkable in itself, it is fortunate in its commanding position, in the woods which form a background to it, and in the sloping lawns which stretch from its door to the Wharfe. Not in sight, but continually in our thoughts, is something better than Denton or Middleton. Beyond the shoulder of Beamsley Fell is Bolton Abbey, the pride of Wharfedale, and it gives a vague charm to the westward prospect to know that the river which shines in our valley at sundown has flowed beneath Barden Tower and through Bolton Woods.

Phi and Theta.

We keep a dog and a cat. The dog is a fox-terrier, the cat an Angora, but neither is well-bred. A really valuable animal runs too great risk of being stolen, and we prefer pets who give us no anxiety. The money value of our animals is negligible, but they are very dear to us.

Fi is a natural abbreviation of Fido, and Fi (spelt Phi) suggests Theta to any one who has been through the elements of trigonometry. It was a professor of mathematics who suggested Theta as a good name for the companion of Phi.

Phi and Theta are very good friends. They will lie down together, and keep one another warm. It is true that when Phi is boisterous, Theta will jump upon a chair or work-table, and if seriously alarmed will spit and strike. She will now and then provoke the dog out of mere caprice. When Phi is trotting past bent upon business of his own, she has been seen to put out a paw and scratch him. But such interruptions of amity are rare.

When Theta first arrived, a mere kitten, and found

herself alone among strangers, and in the presence of a rather terrible dog, she summoned up her courage and stood bravely on the defensive. Nature has taught the cat how to make up for small size and weakness, by pluck and nimbleness. In those early days Phi learned that a cat's claws are sharp and a dog's nose tender.

In presence of a dangerous animal the cat arches her back and erects her tail. This increases her apparent size, and is calculated to strike terror into the enemy. It is curious that the cat assumes very nearly the same attitude when she comes into the room, or sees the family enter. Now her object is to attract attention, and with this end in view, she rubs herself against your legs or your chair, and purrs. The similarity of the attitude assumed under such extremely different circumstances may be explained by supposing that apparent increase of size is useful, either to inspire terror or to attract friendly notice. It seems to me a little discordant with Darwin's Principle of Antithesis, which is, that gestures appropriate to a strongly marked state of mind will be reversed when the state of mind is reversed, and this whether the reversed gestures are serviceable or not.[1]

Where is Angora, and what is the history of the Angora cat? It is a proof of a want of curiosity about certain kinds of facts that very few people can ever tell where Angora is, though the name is so familiar. Put the question in your own family-circle, and see how many know. Angora is in the central highlands

[1] *Expression of the Emotions*, Ch. II.

of Asia Minor, and has been a place of note for 2,000 years. Under the name of Ancyra it was the capital of the Roman province of Galatia. Not only the cats, but the goats and dogs of Angora have thick, long, and silky hair. This is attributed to the action of the climate, which is very cold in winter and hot in summer, and we are told that all these animals lose much of their beauty when taken away from their native country.

I am fond of a cat, and I admire her yet more than I love her. The cat has a beauty which comes of perfect adaptation to a life of emergencies. She is light, swift, adroit, quick to perceive, quick to act. She is most at home on trees, where her wild progenitors sought their prey. The stealthy and self-effacing movement by which a cat in pursu t of a bird creeps along a bough reminds us of a tree-snake. The peculiar iris of a cat, which can change the pupil from a vertical slit to a circle, is excellently suited to an animal which has continually to pass from the shade of dense foliage into full sunlight, and back again into shade. The presence of mind of the cat is marvellous. I have seen a cat chased by two dogs into a corner of a yard with high walls, but the cat escaped unharmed by a gymnastic feat which involved running for several feet up a vertical wall, turning in the air, alighting on the back of one of the dogs, and springing thence to the top of a gate.

Perhaps no animal surpasses the cat, and few rival her in the power of alighting on her feet when accidentally falling. The mechanics of this wonderful accomplishment, which must often preserve the cat's life, have been elucidated by the instantaneous and successive photographs of Marey.[1] One serious difficulty in the way of any explanation consists in the circumstance that a fulcrum, or point of resistance, is required for any movement of rotation, and it is not easy to see what fulcrum the falling cat can employ. It had been previously conjectured that the cat, at the moment of letting go its hold, might use the supporting object as a fulcrum in order to rotate its body as required. This was never a very likely explanation, considering that the cat is usually dislodged by surprise, and that the rotation caused by a sudden shove-off would be pretty sure to continue too long or not long enough. A second hypothesis attributed the turning of the body of the cat to the resistance of the air, but this is disposed of, like the first, by examination of the photographs. The cat, while falling, brings the feet round towards the ground, first the fore feet, and then the hind feet. This is accomplished by the twisting of the body. In order to twist, one end of the body must be fixed, or at least retarded in its revolution. When suddenly let go, the cat gathers up her fore legs, pressing them against her neck, and as near as possible to the axis of the body. In this position their moment of inertia is a minimum, that is, they are as free as possible to rotate. But the hind legs are extended, so as to make their moment of inertia a maximum, that is, to oppose the strongest possible resistance to rotation. The hind legs become for an instant a fulcrum, or

[1] *Comptes Rendus*, CIX., p. 714 (1894). The figures are reproduced in *Nature*, Nov. 22, 1894.

relatively fixed point against which the fore legs can act. Then the fore legs are extended and the hind legs gathered up close to the body, when the inertia of the fore legs furnishes a fulcrum for the rotation of the hind ones. The cat can right herself, and alight on her feet in a very short space. A cat, let go back downwards, with only six inches of clear space beneath her, alighted on her feet. Those who are inclined to repeat the experiment may be warned that the cat dislikes the operation extremely, and that repeated trials are apt to cause vomiting.

A cat will show something which looks like affection. But I fear she is utterly selfish at heart, even when she is happy, even when she affects to love you. Theta is quite demonstrative before meals, and bids you notice how much she enjoys your company. But when the meat is carried out, Theta follows it to the kitchen. She never tries to ingratiate herself when she has been fed within an hour or two. Chamfort detects selfishness in her very gestures,—
" Il ne vous caresse pas ; il se caresse sur vous."

I have set down my general impression of cat nature. But I make haste to add that not every cat is hopelessly selfish. The kindness of the mistress now and then meets with an affectionate return, and the maternal instinct has been known to incline the cat to love an animal of a different species. Mr. Hammond, to whom the readers of my books owe so many excellent figures, tells me that a puppy was brought into his house just at the time when the cat had been robbed of her kittens. She bestowed upon him some of her maternal regard, and to this day will carry about a piece of meat, lay it before him, and gladly see him devour it. Several times she has brought him live mice, in the vain hope that he too would become a mouser.

The dog is simple-minded and has little artifice. Phi will leave his dinner to follow my youngest boy, who delights to race him over the moors. When the snow lay on the hills the boy had his sledge out, and Phi would scamper by his side and bite his feet in full career. He was glad to ride by his master's side down the slopes. Even when put on the sledge by himself, he submitted and shot down the hill-side without flinching, though his attitude was dejected, and he wore an anxious air. I fancy he enjoyed it as little as a man subject to sea-sickness enjoys a sail with a fresh breeze, but he never shirked. If you make a companion of your dog, he will share all risks with you.

The dog attends to what you say; the cat does not, unless indeed she grows in time to understand a particular word as a call to meat. The dog has the idea of conversation, though articulate speech has been denied him. It is well for us and for him that he cannot speak, for I am certain that he would say the same thing over and over again to our utter weariness. Since he cannot speak, he looks at us, and there is great expression in a dog's look, which we should fail to appreciate if he were able to accompany it by foolish speeches.

Society has made the disposition of the dog, want of society the disposition of the cat. The wild dog hunts in packs, and that means combination, some degree of fellow-feeling, some degree of self-sacrifice. But the cat is solitary, goes her own way in silence, and seeks her prey unaided. The short-lived but intense love of the mother-cat for her young ones is the only generous sentiment in cat-life.

How curious that an animal so selfish, so cruel, so fond of concealment and loneliness, should have ever established itself in the dwellings of man! Other carnivores of like tastes have done the same thing. The white-breasted Martin has been supposed to have been the common domestic vermin-killer of the ancients. The Genet is still domesticated here and there on the shores of the Mediterranean, and makes a tolerable cat.[1] Love of mice, it would appear, may in these animals overpower the fear of man. But I suspect that these cats, feline, musteline, or viverrine, were first brought into the house as helpless kittens, and had no choice in the matter. Their usefulness and cleanliness made them agreeable inmates, and the cat for her part came to value shelter, warmth, and food. But she is not truly of the human family; she is a wild animal, which pays us the compliment of residence with us. Her attachment is to the house

[1] Rolleston, " Domestic Cats, Ancient and Modern," *Journal of Anatomy*, Vol. II. p. 57. Rolleston and Hehn believe that no domestic cat was known to the Greeks and Romans. Some of the Greek vases in the British Museum, especially F 207, F 126, and E 171, show cat-like animals which appear to be tame and companionable. The spotted cat led in a string (E 172) is perhaps a Leopard or Panther, which was familiar to the Greeks, as a well-known passage of the *Iliad* shows (XXI. 572-8). The domestication of the cat in Egypt must surely have led to its occasional introduction into Greece and Italy.

and not to us. The cat never longs to talk to us. So little altered is the cat by long domestication that she can manage perfectly well by herself, procuring her own food and bringing up her young in the woods. I have examined a dead cat which had lurked in a copse for about a year, without ever approaching a dwelling. At last it took to felony, stole chickens, and had to be shot. It was sleek and well-nourished, more muscular than common.

The history of our breed of domestic cats is obscure. Naturalists are agreed that it is not identical with the wild cat of Northern Europe. Cuvier could discover no anatomical difference between mummied Egyptian cats and our tame cats, but in this family the specific distinctions are sometimes very slight. Cats have been domesticated in India from remote times. The first mention of the cat in English literature that I have been able to discover is in *Piers Ploughman* :—

"There was no ratton of the rout, for all the reame of France,
That durste bind the bell about the catte's neck." [1]

It has been said that the cat, like sugar and many other useful articles, was first brought to Western Europe in the ships of returning Crusaders. There are, however, indications that some domestic cat, whether of eastern or native origin, was familiar in these islands before the Crusades. The code of Howell dda, published with a translation by the

[1] The *Gesta Romanorum*, which Oesterley supposes to have been written in England towards the end of the thirteenth century, and therefore about a hundred years before Langland, mentions the same fable.

Record Commissioners, dates from the early part of the tenth century. Some, if not all, of the following extracts, probably belong to that time:—

"The worth of a cat and her qualities, this is. 1. The worth of a kitten, from the night it is kittened until it shall open its eyes, is a legal penny. 2. And from that time, until it shall kill mice, two legal pence 3. And after it shall kill mice, four legal pence; and so it always remains. 4. Her qualities are to see, to hear, to kill mice, to have her claws entire, to rear and not to devour her kittens; and if she be bought, and be deficient in any one of those qualities, let one-third of her worth be returned."

The worth of a cat was, according to an old Welsh law, to be estimated thus: the cat was to be held by the end of the tail, with her nose touching an even floor. Wheat was then to be poured over her, until the end of her tail was hidden. "And afterwards this was altered, and there was fixed upon her four legal pence."

"Three animals which reach their worth at a year: a sheep, a cat, and a cur."

"This is the complement of a lawful hamlet: nine buildings, and one plough, and one kiln, and one churn, and one cat, and one cock, and one bull, and one herdman."

"Whoever shall sell a cat, is to answer for her not going a-caterwauling every moon; and that she have ears, eyes, teeth and nails; and being a good mouser."[1]

[1] *Ancient Laws and Institutes of Wales*, pp. 136, 283, 426, 495, 743.

Cormac's Irish Glossary of the tenth century, says that poets eat the flesh of the pig, dog, and cat.

The English law, until a quite recent date, treated cats and dogs as wild animals incapable of being stolen. Even now there are slight differences in the eye of the law between a cat or a dog and a horse or a cow. It is not a felony to steal a dog for the first time, and accordingly the thief is often indicted for stealing the dog's collar, as that entails a more severe punishment.

The status of the dog in some eastern countries suggests that he first entered dwellings for refuse and scraps, that he was a thief and a parasite long before he became the companion of man. But the social qualities of the dog originated still earlier, and were developed by life in the pack. If this is true of the dog, it ought to be true of the wolf too. Perhaps the wolf has social possibilities, and might be made into a delightful companion if only we could get over the awkwardness of the first approaches.

Which are the Wettest Months?

The old name, "February fill-dyke," seems to point to long experience of February as a particularly wet month. On questioning my friends as to their impressions, I am told that the winter months are considered decidedly the wettest, late spring and early summer the driest. But we need not trust to impressions; let us compare our impressions with the rain-gauge. Messrs. Richardson and Co. of York give us the figures for central Yorkshire in their handbook on Artificial Fertilisers. They have taken the average monthly rainfall for forty-six years (1849 to 1894), and have arranged the months in order of dryness, allowing for the varying number of days in the month. " This brings out the interesting fact that in this district the first four months of the year differ but little in their rainfall, and are the driest of the twelve ; that the amount of the rainfall rises steadily through May and June to July, which has usually a much larger rainfall than any earlier month. After this the record declines evenly through August and September, only to rise again in October, which shares with July the joint distinction of being the wettest of the twelve ; the comparatively dry months of November and December leading up to the still drier months of the opening year."

The next thing is to find out whether the distribution of the rain according to season is the same in other parts of the country. Mr. R. H. Scott[1] gives the monthly rainfall for London. March is the driest month, October the wettest, a secondary maximum occurring in July. The seasonal distribution in London is therefore much the same as at York. But when we examine the records of the western counties we find a difference. The maximum for the year comes later, in November, and the spring is not so dry as in the eastern counties.

Are our beliefs as to the wetness of winter and the dryness of summer mere mistakes ? Not at all ! We judge by the state of the ground, not by the amount

[1] *Elementary Meteorology*, Fig. 38.

of rainfall. In the cold season the evaporation is greatly less than in summer. Hence though less rain falls in December and January than in July and August, it fills the dykes much more effectually. The ground is wetter, the springs and rivers fuller in average winter than in average summer weather.

ANIMALS WITH AND WITHOUT COMBS.

I sit by the fire and lazily watch Theta cleaning and smoothing her fur. She not only washes, but combs her fur with her tongue. We have all allowed some pet Cat to lick our hands, and know very well that she has a rough tongue. Cuvier tells us that the Lion's tongue is so rough that it can be used to rasp the flesh from the bones, and it has been said that the Cat's tongue is used in the same way. In the case of the Lion, the horny, recurved, claw-like papillæ are nearly a quarter of an inch long, but I doubt whether the Cat's tongue is an efficient rasp. What then is the use of the horny papillæ which the Cat too possesses? I think that they are chiefly serviceable as a comb, and that it is because the Cat bears fur, and not because she devours flesh that she has a prickly tongue? Are then all fur-bearing animals provided with a prickly tongue? By no means. There are other ways in which fur can be kept sleek besides combing. The Rabbit, for instance, washes his face like a Cat, but there are no prickles on his tongue. How he keeps his fur in good order I do not know. The Fur-seals would, I imagine, find it an unpleasant task to lick their vast bodies all dripping with salt water. But the long, coarse and deep-rooted bristles which lie in the fur keep it from getting matted or ruffled. Some animals can use their claws as combs; in others the pile of the fur is too short to need combing at all.

The Cat does her licking by preference after a meal, probably because the saliva flows most freely at that time. Then she likes to go to sleep. The three actions of feeding, licking and sleeping have become associated, not only in the Cat's memory, but very likely (so uniform is the practice) in the nervous mechanism of her body. Some men associate feeding, smoking and sleeping, but this is merely the habit of an individual, and not ingrained in the physical organisation of the race. There are men who eat without wanting either to smoke or sleep, and many women put knitting in the place of smoking. But every Cat that I have known loves to lick after eating, and to sleep after licking.

Many Birds possess a useful comb in the claw of the middle toe of the foot, this has been noticed in Owls, Night-jars, Herons, Bitterns, Cormorants, Gannets, etc. It has been explained as a means of holding the prey securely. Gilbert White probably set this notion afoot. In his forty-seventh letter he says of the Goatsucker or Nightjar:—" I saw it distinctly more than once put out its short leg while on the wing, and by a bend of the head deliver somewhat into its mouth. If it takes any part of its prey in its foot, as I have now the greatest reason to suppose it does these chafers, I no longer wonder at the use of its middle toe, which is curiously furnished with a

ANIMALS WITH AND WITHOUT COMBS 51

serrated claw." Mr. E. B. Titchener[1] holds that this explanation cannot be right, first, because the serration is small in extent; and secondly, because it is at the side and not on the under surface of the claw. The Goatsucker is said to clean its mouth-bristles with its middle toe, but the mouth-bristles and the comb do not always co-exist in the same species. A young Heron was kept under observation to see how it employed its claws. Its food, whether living or dead, and whether taken from water or from the ground, was never touched at all by the feet. The only use to which the serrated claw was put was that of scratching the cheeks and throat. In this action, which occurred most frequently after a meal, the other two front toes were curved down, so as to leave the middle claw free. Mr. H. R. Davies[2] confirms Mr. Titchener's view by some fresh observations. A Cormorant was found to have the fissures between the teeth of its serrated claw choked with fragments of down, corresponding with that on the body of the same Bird. Minute fragments of feather were afterwards found in the claw of a Barn Owl. The comb is sometimes replaced by a curved blade with teeth, which runs along the inner side of the claw. Such

FIG. 11.—Claw of Heron, magnified.

FIG. 12.—Claw of Nightjar, magnified.

[1] *Nature*, Dec. 4, 1890.   [2] *Nature*, Feb. 19, 1891.

a blade is found in Guillemots, Razor-bills, Wild Duck, Teal, Gulls, Oyster-catchers, Golden Plovers, Starlings, Fieldfares, Redwings, Larks and many others. In Divers, Partridges and Pheasants the claw is flattened so that its inner edge forms a scraper. Where a comb is required the inner

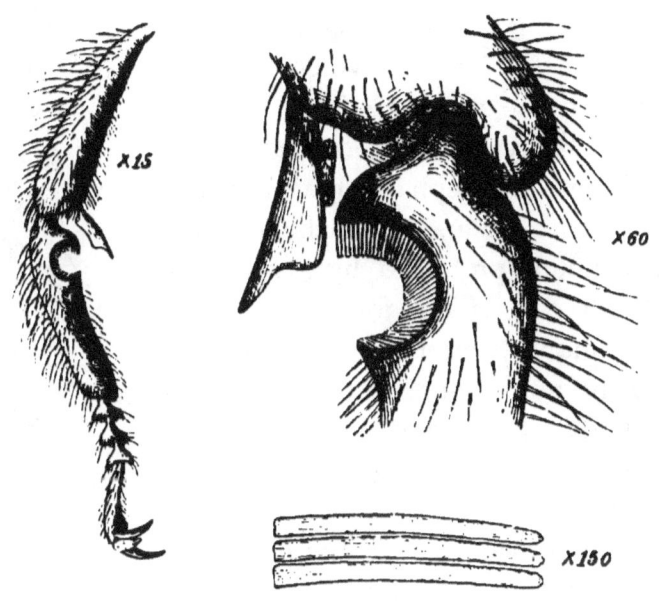

FIG. 13.—Fore-leg of Bee with tibial comb; the comb more highly magnified; three teeth of the comb.

edge of this blade becomes divided into teeth. Young Nightjars have only the blade, but old ones have a well-developed comb. Mr. Titchener [1] adds that Audubon once shot a Frigate-bird, and found the comb crammed with the Insects which occur on the head and especially about the ears

[1] *Nature*, Feb. 19, 1891

ANIMALS WITH AND WITHOUT COMBS 53

of the Bird. Hudson[1] is quoted for the observation that Herons are remarkably free from vermin, while the Roseate Spoonbill is infested by them; both have the serrated claw. The Herons (captive?) were always in a miserable condition; the Spoonbills plump and healthy.

Fig. 14.—Part of fore-leg of Aëpus, a marine Beetle, with comb upon the tibia.

The Honey-bee has a comb in the fore-leg, lying in the angle between the tibia and tarsus, which is used to cleanse the antennæ from dust or pollen. Many Beetles, belonging to the section Geodephaga, have a comb of like structure and use, which forms a deep notch, protected by a spine, at the lower end of the fore tibia.

The mouth, whether armed with a comb or not, is often used to keep the body trim and clean. Cockroaches draw their long antennæ from time to time through the mouth. Simulium larvæ cleanse their fan-like brushes in the same way. The larva of the Gnat may be seen busily at work clearing its body of attached Infusoria, and devouring all that it can reach.

There are some animals, such as bivalve Mollusks which have no effective means of removing foreign bodies from their skin. They sometimes find it the simplest plan to coat the irritating object with nacreous shell, as the pearl-forming Hyria does with the images thrust under its mantle by the crafty

[1] *Argentine Ornithology*, Vol. II.

Chinese. How it must plague the Pond-mussel to be overrun, as it almost always is, with Water-mites, and yet be unable even to scratch itself! The Pinna harbours a small crab, the Pea-crab, within its shell, and must, one would think, wince at times as the hard, pointed legs press against its unprotected flesh.[1] But use is everything. Dogs are said to turn melancholy if they are kept absolutely clear of Fleas, and perhaps the Mollusks, whose sedentary life in a dark shell must be dulness itself, find their parasites a source of mild excitement.

The Moon.

Townsfolk do not care very much about the Moon, nor observe her very carefully. Now and then they admire the slender crescent in the western sky, or the full Moon flooding the landscape with her cold light, but the Moon is of little practical use to people who live in well-lit streets. It is quite different in the country, where there are no gas-lamps. If there is a Moon, we travel comfortably along, with light enough to steer our course; if there is none, we are liable to step off the foot-path into a puddle, or to walk into a bush. Country people often arrange their meetings so as to walk home by moonlight, and not on nights when there is no Moon.

Even if the sky is hidden by clouds, the Moon, especially when half or more than half-full, can send a faint, diffused light through the clouds. It is seldom

[1] Another species of the Pea-crab is not uncommon within the valves of the edible Mussel.

quite dark, dark as a pocket, except when the Moon is below the horizon and the stars are clouded.

It makes a great practical difference to us whether the Moon rises early or late. If she does not appear till we have got home and gone to bed, she might as well never appear at all. Any observant person living in the country will soon find out that two or three days after new Moon, she is to be seen in the evening, being then near to setting; that the full Moon rises about sunset and shines throughout the night, setting at sunrise; and that the waning Moon rises later and later every night until she rises in the early morning, setting a little before sunset. We commonly see the Moon at night, whenever the sky is clear and the Moon from three to twenty days old. After that we see little of her at night until after the new Moon. But even at times when we do not see the Moon between sunset and midnight, we can often see her in the day-time. In the earlier part of her last quarter, the Moon rises in the morning and sets in the afternoon. The Moon in her first quarter generally rises in the afternoon, and sets early in the evening. At these times we may expect to see the *day-Moon*, if we look out for her.

I shall take the liberty of explaining many things which everybody is supposed to know, but I think I need not explain the causes of new and full Moon. Taking so much for granted, I will mention a fact which, as I find by experience, is not known to all intelligent and well-read people. We can tell by looking at the moon whether she is waxing or waning. If she is waxing, the illuminated edge is to our right hand; if she is waning, it is to our left hand. The reason of this is easily seen if we pass a ball round the head in a room lit by a lamp or a single window. If the ball is made to circle *with clock-hands*, the dark side will travel foremost, and will lie to the right hand, while the ball recedes from the light. During this part of its course the ball will appear more and more illuminated as it moves. When the ball approaches the light, the bright side will travel foremost, and will lie to the right hand. If we reverse the direction of the ball, the phases will succeed one another as they do in the Moon. It is evident from this that the Moon circles *against clock-hands*.

The bright side of the Moon is always turned towards the sun, whether the sun is visible to us or not. Hence we should expect that, if the Moon is less than full, a line joining the centre of the Moon and the centre of her illuminated edge would always point towards the sun, while the *cusps* or horns of the Moon in her first or last quarters would point away from the sun. But if we come to watch the Moon we shall find that the position of her cusps is often different from what we had expected. For instance, the sun may be well below the horizon, yet the cusps may be turned a little downwards, and the centre of the bright edge a little upwards. Some fanciful people have even supposed that the position of the Moon's cusps varies according to the weather which we are going to have! With or without reason persons of strong imagination are always hoping to find some connection between the Moon and the weather.[1]

The rule which governs the position of the Moon's cusps and bright edge is easily stated, though it is not so easily explained. Imagine a great circle drawn across the star-sphere, passing through the centres of both sun and Moon. That circle will nearly coincide with the zodiac, and it may be taken as the edge of a circular plane in which the earth lies. The centre of the bright edge of the Moon and the centre of the line joining her cusps will be found always to lie in that great circle; the bright edge being turned towards the sun.

When the Moon is almost new, she presents a bright crescent, enclosing a faintly illuminated surface, which we call the new Moon in the arms of the old. The light thus faintly reflected cannot come direct from the sun, nor is the Moon self-luminous; it is reflected from the earth. At new Moon the Moon is between the sun and the earth, and the earth appears full to the Moon, reflecting its maximum of sunlight upon the Moon. At half-moon the "earth-shine" is much fainter, for then only half the earth's disc, as seen from the Moon, is illuminated.

What is the apparent size of the Moon? Very nearly the same as the apparent size of the sun. At eclipses of the sun, we see that the Moon very nearly

---

[1] In some parts of the country the belief is that when the Moon holds the mouth of her cup uppermost it will be fine, but that it will rain if the cup seems to be turned upside down.

or quite covers the sun's disc. We cannot give the apparent diameter of the Moon in any measure except angular measure. It is about $\frac{1}{2}°$. 720 such Moons would make a belt going all round the horizon. 360 such Moons would make an arch passing through the zenith from horizon to horizon.

There is no comparison between the light of the sun and that of the Moon. If every part of the sky were as bright as a full Moon we should not receive as much light as in full day-light. Wollaston estimated that the sun gives out 800,000 times as much light as the full moon.

The path of the Moon among the constellations can be observed by any one who will take a little trouble. Her motion is sufficiently rapid to cause her place to change visibly in a few hours. How rapid is it? The Moon completes the circle of the heavens in $27\frac{1}{3}$ days. She therefore travels about $13°$ daily on an average, and a little more than her own diameter in an hour. The shifting of the Moon from night to night can be followed in clear weather by making a plan of the constellations near her path, and noting upon it the place of the Moon every evening.

By carefully noting the Moon's path among the stars, it has been found out:—

1. That she keeps very near the zodiac, never departing much more than $5°$ from the ecliptic, or apparent path of the sun.

2. That she does not take exactly the same path every time, and does not end exactly where she

THE MOON

began. It takes $18\frac{1}{2}$ years before she travels over the same path a second time.

3. That she travels in the same direction as the sun and the chief planets. That direction is *against clock hands*, contrary to the direction of rotation of the star-sphere.

4. That she completes her circle of the heavens in 27 days, 8 hours, or $27\frac{1}{3}$ days.

Since the Moon completes her round in $27\frac{1}{3}$ days, it would seem to follow that we shall have a new Moon every $27\frac{1}{3}$ days. But this is not the case. We have a new Moon only when the Moon is as near as possible to the sun, on the same meridian as the sun, and it takes more than $27\frac{1}{3}$ days to bring her round to that point. For during all the time that the Moon is travelling round the ecliptic, the sun is travelling too, and in the same direction, though much slower. When the Moon has completed her circle, she has still to go over the distance travelled by the sun since the last new Moon. How much is that? The sun (apparently) travels round the heavens in a year. He will, therefore, travel $\frac{360°}{365}$ or about a degree a day. That is about 28° in 28 days. At the Moon's average rate of $12\frac{1}{2}°$ per day, it will take her a little over two days to travel those 28°, and to overtake the sun, after she has completed her revolution round the earth. Hence the interval between one new Moon and the next is $29\frac{1}{2}$ days, while the revolution of the Moon round the earth occupies only $27\frac{1}{3}$ days.

What is the figure of the Moon's path in space? If we could look at the Moon from a very great distance, much greater than the diameter of the earth's orbit, and if our point of view were in the line joining the North and South Pole of the heavens, the Moon might be seen to circle round the earth, while both would circle round the sun. By careful observation of the place of the Moon in the background of stars her path in space could be mapped with any degree of precision that might be desired.

We should find if the trial could be made that the path of the Moon in space is so nearly identical with the earth's orbit that very close observation would be required to distinguish them. In its course round the sun the Moon would make thirteen very gentle undulations, curving outwards from the earth's orbit for a very trifling distance thirteen times, and curving a little within it as often. But it would require a very large sheet of paper and very careful drawing to make the difference apparent, for the deviation from the earth's orbit would not at most exceed about $\frac{1}{4}$ per cent., that is $\frac{1}{4}$ in. in 100 inches. (8 ft., 4 inches.)

The times of rising and setting of the Moon are influenced by the same causes which affect the times of rising and setting of the sun, but not quite in the same way. The Moon travels nearly along the zodiac. The new Moon must rise and set nearly at the same time as the sun, because she is near to him. The full Moon will rise at about the same time that the sun sets, and set at about the time that he rises, because she is then opposite to him in the zodiac. Therefore in winter, when the sun rises late and sets early, the new Moon will do the same, but the full Moon will rise early and set late. In summer the case will be just reversed. The new Moon is longer above the horizon in summer than in winter; the full Moon is longer above the horizon in winter than in summer.

The Moon's face exhibits a pattern which is always much the same, for the Moon turns the same face towards the earth. There is, however, a belt about the Moon, occupying about one-fifth of her surface, which is sometimes visible from the earth and sometimes not. The pattern on the Moon was formerly supposed to be due to continents and seas, but telescopic examination shows that the Moon has no water on her surface. More than this, she has either no atmosphere at all, or an atmosphere very much less dense than that of the Earth. When she passes between us and a star, the star disappears suddenly, instead of being lost in haze. No clouds are even seen to hide the pattern of the Moon's surface.

The telescope reveals the cause of the peculiar markings. Lord Rosse's great telescope magnifies 6,000 diameters, and should have the effect, apart from disturbing causes, of showing us the Moon at a distance of only thirty-nine miles. But no telescope is optically perfect, and the earth's atmosphere greatly interferes with clear vision. We do not get nearly so good a view of the Moon as the magnifying power of the largest telescopes would lead us to expect. The patches and dots upon the Moon are resolved by a good telescope into crater-like mountains, casting very sharp shadows. Some of these craters are very large. Tycho has a diameter of more than fifty miles, and covers an area almost as large as the West Riding of Yorkshire. They are commonly believed to be true volcanic craters, but some of them are unlike any terrestrial craters. There are deep circular depressions, ringed about by steep walls, plains ringed in the same way, craters with central elevations, and circles of craters, as well as small craters of the usual terrestrial form.

The fact that the Moon turns always the same face towards the earth implies that during every revolution round the earth she rotates upon her own axis. If a man walks round a tree, always keeping his face towards it, he will face all quarters of the sky in turn, that is, he will rotate. He will rotate in the same direction as that in which he walks round the tree. Thus the Moon rotates once in $27\frac{1}{3}$ of our days, and rotates against clock-hands.

If we could stand on the surface of the Moon, we should see great and small craters, some towering into the sky, others low on the horizon. The distant ones would be clear and sharp, for there are no clouds or haze. We should see no streams or pools, no long, narrow, branching valleys, and probably no water-worn rocks or rounded pebbles. Some observers, however, think that they can detect upon the Moon traces of the former action of water and even of moving ice.

Let us suppose ourselves planted on that side of the Moon which is turned towards the earth, and that the sun shines upon us. The irregular surface of the ground is lit by a fierce light, and all objects cast deeper shadows than are ever seen on earth. Though it is day, the sky is black and the stars shine with intense brilliancy. The tender blue of the earthly sky has vanished; it was entirely due to water-vapour, and there is no water here. The sun is too dazzling to behold, and his heat so scorching that no human being could endure it unscreened. He seems of immense size, for around his disc is a glory of extraordinary brightness and great extent, which flashes like an aurora borealis. The earth hangs in the sky as a vast disc which goes through its phases and appears dark, partly dark, or wholly bright at different times. She is thirteen times as large as the full Moon seen from the earth, and is almost a fixed object in the lunar heavens. Her edge is fringed with a narrow luminous cloud. The day with its insupportably bright sunlight lasts $27\frac{1}{3}$ earth-days; the year is about as long as an earth-year. Day and night are nearly equal throughout the year, but the height of the noon-day sun varies according to latitude and season, just as it does upon earth.

It is a help toward understanding eclipses and other lunar phenomena to form a true mental picture of the sun, moon, and earth. Let the earth be a large marble, one inch in diameter; then the Moon will be a small pea at a distance of $2\frac{1}{2}$ feet, and the sun a 9 foot globe at a distance of 320 yards. A hollow globe as big as the sun, with the earth in the centre, would give ample room for the Moon to revolve at her usual distance.

If the position of the sun, earth and Moon with respect to one another are noted at a particular date, it will be found that in about eighteen years they will again occupy very nearly the same position. Accordingly eclipses, both of the sun and Moon may be expected to recur after this interval of time. But the correspondence is never quite exact, and the rule is an imperfect one. In old times, before the motions of the Moon were thoroughly understood, there was no better way of predicting eclipses than this, and it was found to work tolerably well.

Spring Crocuses.

The Snowdrop heads the procession of spring flowers. Then comes the Crocus, and a little later, the Hyacinth. The Narcissi follow, and keep us gay till early summer, when the gardener has neither space nor leisure for all the things that are ready to come into bloom at once.

What tempts the Crocus to flower so early, before the snow has quite gone, and when night-frosts may be expected for two months yet? It must be an advantage to the plant that its flowers appear before the grass begins to grow, and its attractiveness to the Insects which emerge so early will be unusually strong. The autumn-flowering Crocuses enjoy a like advantage. Both find it hard to bring their wares to market, and there are few customers; but then there is little or no competition among the dealers.

Can we be sure that the Crocus is insect-fertilised? Its bright colours and large size testify to its need of attracting the notice of animals, and the slender tube of the flower is filled to the brim with honey. Hermann Müller can tell you what Insects fertilise the purple

Spring Crocus (*C. vernus*). He has seen a common night-moth (*Plusia gamma*) and the Painted Lady Butterfly and the Humble-bee visiting the flowers. The anthers open first and shed their pollen; afterwards the stigmas ripen and expand. If good-sized Insects are attracted to the flowers, it is well, for then the pollen will be laid upon the stigmas of another plant; but failing this, the stigmas curve downwards upon the anthers, and get dusted; fertilised, but not cross-fertilised.

Our common yellow spring Crocus is *C. aureus*, a native of Turkey, Greece and Asia Minor. It is known by its short and little-branched stigmas, and its suddenly diverging anthers. *Crocus vernus*, the purple or white spring Crocus of the gardens, is a native of the Alps. Its stigmas are of a deep orange colour, and contrast strongly with the rest of the flower.

The seeds of the Crocus ripen at midsummer, and should be sown at once if it is intended to raise bulbs from them. The plant raised from seed is not ready to flower for two or three/years. Nearly all our Crocus bulbs are grown in Holland and Lincolnshire.

Plants which bloom very early or very late in the year, do so at the expense of food laid up in the previous summer. Hence they are often bulbous, containing much starch or sugar in the coats of the bulb, which are either future leaves, or the bases of old ones. Sometimes they have tuberous roots, like Cyclamen, or a perennial, woody stem, like the Mezereon. Annual plants, with thin, fibrous roots and little wood, can hardly flower except in seasons when the bright sun makes food-formation easy.

Let us take an ungerminated bulb of last year's growth, and examine it. It has a circular scar at its base, round which the roots spring, when there are

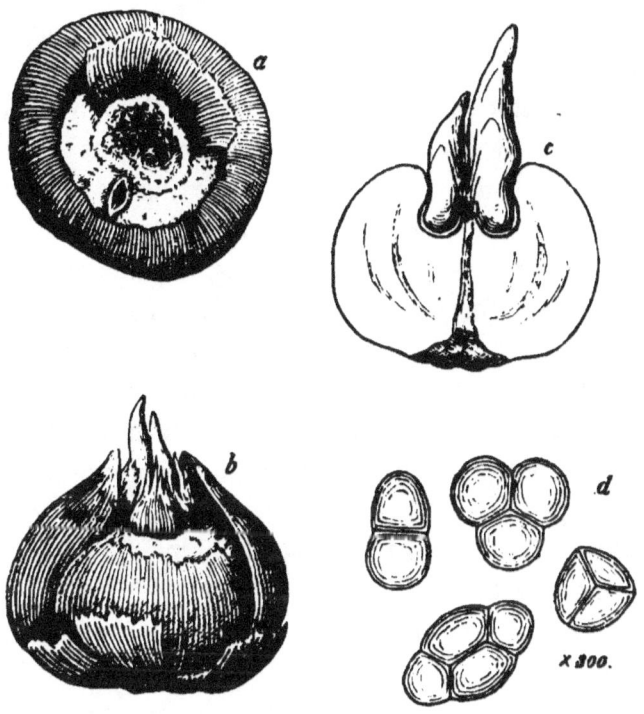

FIG. 15.—A crocus corm. *a*, the base; *b*. side view; *c*, section, showing two shoots; *d*, starch granules. *a—c* are of the natural size; *d* is magnified.

any. Above this, the bulb is covered by an outer tunic, consisting of a membrane stiffened by many prominent, vertical fibres, with frequent junctions. When such a membrane shrivels or rots in the earth, it becomes reduced to a network with a ragged fringe, such as we have now before us. It is firmly attached to the circular scar, but free above. Now and then the outer tunic is double, those of two seasons persisting together.

If we strip off the outer tunic, we expose a white mass, not formed of overlapping coats as in a Hyacinth or Onion bulb, but solid and nearly uniform in texture. From this, and usually from its top, spring the shoots, which will rise into the air, bearing leaves and flowers. There are no roots at present; the old roots are withered and the new ones do not form until the shoots begin to push. Surrounding the shoots are several other tunics, of the same texture as the outer one, but of smaller size. They are like circular capes, laid one above another to protect the tender shoots. At first, these inner tunics are all attached at various levels to the white mass, but as the mass enlarges the lowest tunic becomes detached and is slipt upwards, leaving a distinct circle on the surface a little above the basal scar, to show where it was once attached. The remaining tunics often remain in place; if you pull them off they leave similar circular lines or scars.

Since the chief part of the bulb is not made up of scales or coats, it is not in technical language a bulb at all, but a corm. What is the substance of the corm made of? Pare a slice off, and put it into weak iodine solution (tincture of iodine diluted with water). Blue specks immediately appear on the cut surface, and before long it turns blue-black all over. Blue grains wash out into the solution. If these are examined by the microscope, they are found to be rounded and marked with concentric lines. The shape, the markings, and the blue colour with iodine prove that these are grains of starch. In the Crocus corm many of the grains are compound, consisting of several which cohere together. A great part of the corm is composed of cells filled with starch. There is also some sugar and a very little albumen. This food is laid up for the future use of the plant, and may be employed as human food. In Syria, Crocus corms are sold in the markets, roasted and eaten.

Now take a razor or a sharp knife, and slice the corm through the middle, taking care to cut the principal shoot symmetrically. We have now cut through the mass of starchy food, and lying in it, we see two or three greenish or yellow streaks. These are bundles of vessels, many of which, if carefully followed, will be found to pass into the bases of the leaves or tunics. The shoot, when cut through, is seen to be made up of leaves in successive layers. The outer ones are protective merely, and soon wither; then come leaves, which will turn green and form the assimilating organs. Within these are the flower-sheath and the flower itself. The yellow petals and the ovary with its numerous seeds can be made out in a shoot an inch long, and with the help of a lens, in still smaller shoots.

All the parts which we call tunics, leaves and sheath, and not only these, but the sepals, petals, stamens and carpels of the flower, are essentially leaves, various in form and function, but alike in origin.

If the tunics of the corm are leaves, what is the

SPRING CROCUSES 69

corm itself? It bears roots, and leaves, and flowers. Clearly it must be the stem, a stem which is very short, thick and fleshy, but as much a stem as a cabbage-stalk is. You know the prostrate stem of

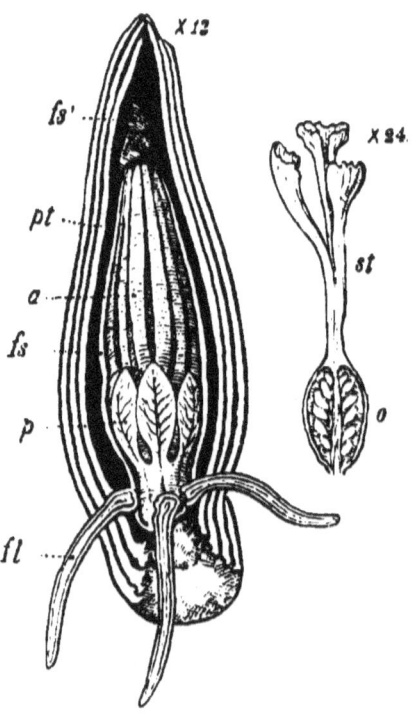

FIG. 16.—Shoot of crocus, laid open to show the flower within. Also the pistil removed from the same flower; *pt*, protective tunics, three in number; *fl*, foliage-leaves, turned down, so as not to interfere with the view of the flower; *fs*, flower-sheath or bract; *fs'*, sheaths of other flowers; *p*, petals; *a*, anthers of the stamens; *st*, style, with three stigmas; *o*, ovary.

an Iris or a Solomon's Seal. Imagine it set upright and reduced in length to a mere button, from which leaf after leaf springs. The flowers are borne on the low, cup-like summit, and there are eyes or buds, as we shall shortly see.

I think it likely that remote progenitors of the Crocus had an upright, fleshy stem, with sheathing leaves and flowers at the top. The necessity of storing up a large supply of food for flowering out of the usual season, seems to have caused the stem to enlarge in width, and diminish in height, until it became a sphere, and even a button. The leaf-bases became crowded together; the lowest disappeared as foliage-leaves, leaving only a web of fibres, such as you may see at the base of a palm-leaf. Such reduced leaves form the tunics or protective layers. Normal foliage-leaves and flowers were given off from the summit of the stem as before. The structure of the Crocus-stem and its history from year to year are peculiar, yet not so peculiar but that we can usefully compare it with plants of a more ordinary kind.

It is well worth while to dissect out all the parts of a young flowering corm. Take a Crocus in flower, separate one of the small, new corms from the old one, and strip off its envelopes one after another. First come the brown and withered tunics, then a number of soft, white sheaths (the new tunics). Next come the foliage-leaves, one enclosing another. Dissect these carefully away from the corm with needles, and observe that each has a white, ring-like scale at its base, which is plainer in the outermost leaf than in the others. Even the strap-like foliage-leaf seems to be derived from a sheathing, tubular leaf. Inside the foliage-leaves come the flowering branches. Each is apparently enclosed by a single whitish sheath or bract, but if this is slit open, it will be found to be double, a narrow pointed bract springing from the inside of the outer tubular one. Both are united to the flower-stalk beneath the ovary.

I will leave you to make out the structure of the Crocus-flower, which the common manuals of Botany will help you to do, if you require help at all. Let us carry the life-history a stage or two further. After flowering, the foliage-leaves remain active for several months, and fill the young corm with food. All parts of the flowers, except the slowly-ripening ovaries, wither away. By the end of June or the beginning of July, the seeds are ripe, the seed vessel raises itself from the ground, opens its valves, and the seeds are dispersed. Then the foliage-leaves turn brown, the roots wither, and the plant enters upon its resting-stage.

There is now no outward sign of change or growth. Hardly anything is taken in or given out for months together, and the corm seems dead. Dead it is not, however, for during this resting-stage, and especially in the earlier part of it, next year's corm is matured. This looks like a new plant, budded out from the old one; but it is really only the enlarged base of a branch—the branch upon which the leaves and flowers of the preceding spring were borne. In summer the leaves and flowers wither, and the branch dies down to the enlarged base. This does not wither, but absorbs the nutritive substance of the old corm, and at length completely replaces it, being provided with a new set of tunics, and later on with a new set of roots.

Other branches may form within the lower leaves,

and produce new corms. In some species of Crocus the old corm produces a considerable number of new ones. If a yellow or blue garden-crocus is planted just below the surface of the ground, it will divide into two or three smaller ones, which only attain their full size after two or more years of growth.

It will clear up your notions about corms and bulbs to take a kitchen-onion, and slice it through the

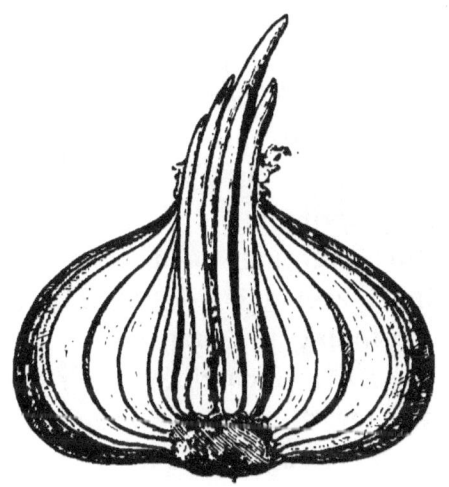

FIG. 17.—Bulb of Onion, in section.

middle. Almost the whole bulb is made up of leaf-bases. In the centre we can make out the undeveloped head of flowers. Towards the base is a fleshy knob from which all the leaves and flowers and roots spring. This is clearly the greatly reduced stem. If you suppose that three or four such bulbs had a large common stem, we should get something like the corm of the Crocus. Or suppose that a Crocus corm had only one large shoot, and that the stem dwindled to an insignificant size, we should get something like the bulb of the Onion.

Lay the cut surface of the Onion-bulb in a saucer of dilute iodine solution. It does not change colour. The Onion contains no starch, but plenty of sugar instead.

CATKINS.[1]

March 16, 1895.—This is the first entirely pleasant day of spring. A soft air, a gentle west wind, continuous though veiled sunshine. The long grass has been turned grey by the hard winter weather of January and February, but close to the ground green tufts are already springing.

Walking this morning in a little copse, I saw catkins on the Hazel and Alder. The buds of the Willow are beginning to part, and to show a silvery gleam from the hairs which clothe their bracts. Snowdrops (much later than usual) are in flower. The Crocus, too, is flowering, but only in favoured spots. The ten weeks' frost has kept them back far beyond their usual time.

Many seeds dispersed by the gales of winter are beginning to germinate. The seedlings of the Sycamore are plentiful, some just pushing out their green radicles from the scar which marks the former adhesion of the seed-vessel to its fellow, others just escaped from the seed-vessel, but still enveloped in the brown seed-coat, others quite free, and beginning

[1] The word (German *Kätzchen*) means *kitten*. In some country places the catkins of the Sallow are called *kittens* and *catstails*.

to unroll their crumpled seed-leaves. The pink seeds of the Elm are to be seen here and there, newly disengaged from the winged fruits.[1] A few Birch-cones lie about the roots of the trees from which they fell, and some still hang on the bough. Some are still full of winged fruits, but most are empty or nearly so. The fruits of the Birch are scattered far and wide over the fields. I found some which were 250 paces from the nearest Birch-tree, and there seems no reason why they should not travel miles through the air in a full gale of wind.

My eyes are not so good as they were now that I am turned fifty, and it is a great help to have quick-sighted boys as companions of my walks. Years ago I trained my boys to observe the common sights of the country, and now I reap the benefit as well as they. The schoolmaster might heap up natural knowledge if he could learn to see with his boys' eyes as well as his own, for the curious school-boy will work over the country like a dog, putting his head into every hole. But too often there is no one to share the boys' little discoveries, no one to give the gentle shove that is wanted at a sticking place. The schoolmaster is of course a learned man, perhaps a divine. As he strolls along he is thinking of a new theory of the Absolute, or of a method in Higher Algebra, or of next Sunday's sermon, or of a disagreeable letter that came by the morning's post. If the schoolmaster cannot stoop, or run, or climb, or tell the notes of the birds, or mark the common flowers (and

[1] The seeds are those of the Witch Elm. The Common Elm never seeds in Britain.

many excellent schoolmasters can do none of these things), it would be well to find an active and observant deputy to join the rambles of the boys. A master in sympathy with the boys and with nature would learn almost as much from the boys as the boys would learn from him.

If you have boys and girls about you, whether your own or other people's, take them into the woods and fields. Try to answer their questions; try to put better questions than they can think of. Never mind the technical names; leave all your Latin and Greek at home. One of the best (and hardest) questions is: —"What is the use of this to the plant or animal?" Do not be discouraged if, as will generally happen, no one can tell. You will grow a little more expert with practice, but to the last you will find many simple-looking questions quite insoluble. Never shrink from saying, "I don't know." These words are always on the lips of a well-trained and inquisitive naturalist. It is dismal, though common enough, to put words in the place of knowledge. "Wo Begriffe fehlen, da stellt ein Wort zur rechten Zeit sich ein."

But I am running to words myself. Let me take up some special thing for examination and report. I will study the Catkins, which will be plentiful for the next fortnight or three weeks. Alder and Hazel catkins are already to be seen everywhere, and the Birch and Willow will be out in a few days.[1]

---

[1] The descriptions which follow will be found dry and pretty nearly unintelligible to those who have not the catkins and cones actually before them. Descriptions cannot stand in place of the things themselves, but are merely a help to the observer.

The catkins of the Alder, like most other catkins, appear before the leaves. The tip of the branch forks beyond the furthest leaf-bud, and each fork divides in

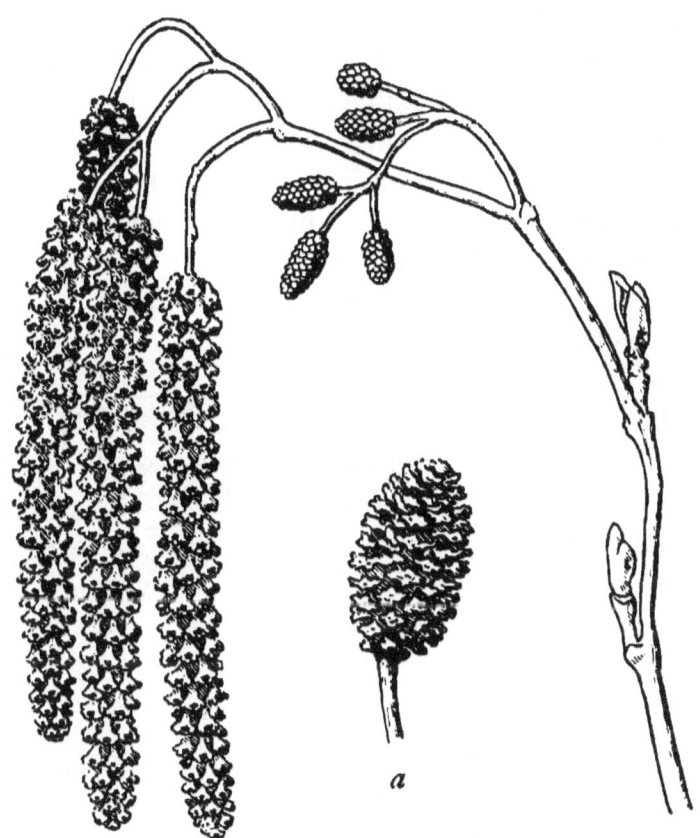

FIG. 18.—Flowering branch of Alder (*Alnus glutinosa*), with catkins and cones. *a*, a ripe cone.

two, three or more branches. All the branches of the same fork carry flowers of the same kind, either male (stamen-bearing) or female (ovule-bearing). It cannot be decided by mere inspection whether this is a case of true forking, or whether one branch, that which bears male flowers, is given off laterally, while the true termination of the stem bears the female flowers. The male flowers are borne upon spikes (catkins) two or even three inches long, which are at first rigid, but afterwards become flexible and droop; the spikes of female flowers (cones) never droop, but tend to become more erect during and after flowering. The flowering spikes were all formed last summer, and could be seen in their unexpanded state at any time during the winter; they are from the first unprotected by envelopes of any kind.

On one of the drooping male catkins we see a great number of scales given off from a central stem. The scales are now parting, and between them the bunches of stamens can be seen. Cut off a single scale and examine it. There is no better way than to impale it upon a pin, thrusting the pin into the base of the scale, which can then be turned any way at pleasure, and the parts studied with a pocket-lens. We see that the scale ends in a shield-like expansion of crimson-purple colour. To its edge are attached two smaller bracts of the same colour, and in the angles between these and the central lobe are two more bracts. The up-

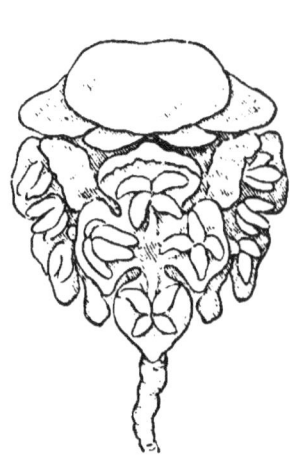

FIG. 19.—Scale of Alder-catkin, with male flowers. Magnified.

turned surface of the scale is bare; in a ripe catkin it is dusted with abundance of pollen shed from the flowers above; from the side which hangs down the flowers spring. It is plain that they are well sheltered from the rain by the over-arching scale.

There are three flowers to every scale. Each is enclosed in a calyx of four sepals, and opposite each sepal is a stamen. In the unexpanded catkin the stamen completely fills the space within the hollow sepal, but as the flowers ripen the stamen becomes free, each of its two large anther-lobes bursts, and the yellow pollen is shed.

FIG. 20.—Two cones of Alder in flower. Magnified.

Now let us take one of the female cones and examine it closely. The scales can be parted with needles, and examined with a lens. Each scale is hollowed out, well rounded sides, and pointed tip. Close to its base two ovaries can be seen, each bearing two styles. The ends of the styles can often be seen protruding between the scales; at the tip of each is the stigma which has to catch the grains of pollen.

Two styles indicate two carpels, and microscopic examination shows that there really are two carpels to each flower. The ovary is at first two-celled, and contains two seeds. But only one of these seeds will become completely developed; the other will be squeezed out of existence by its neighbours, and the ripe ovary will be one-celled and one-seeded. Behind each ovary are two minute bracts, lying side by side against the scale. These can only be made out by close examination; they become plainer during the ripening of the seed.

On the same trees which yield the catkins and cones, last year's cones, black and woody, and perhaps the cones of the year before last, can be seen on the ends of some of the branches. When the cones come to maturity, their scales part and the fruits are exposed. By March many of last year's fruits have been shaken out, but plenty can still be found on the tree. Break a ripe cone across. You will find the central stalk extremely tough. From it radiate the woody scales, each bearing a pair of angular fruits upon its upper surface.

Fig. 21.—Two flowers from a cone of the Alder, showing a large outer scale, and two pairs of smaller bracts, one pair to each flower. Magnified.

We saw that the Birch-fruits were winged; why are not the Alder-fruits winged too? Does this invalidate the explanation that the wing is useful in dispersal? I think not; it is only a *negative exception*. A wooden leg is used to enable a man to walk when he has lost his natural leg. If you saw a one-legged man walking with a pair of crutches, and no wooden leg at all, would that shake your belief in the motive for wearing wooden legs? I shall have something more to say about the dispersal of Alder-seeds by and by.[1]

Birch catkins and cones are not unlike those of the Alder. Each scale of the male catkin has four bracts and three flowers, as in the Alder; but each scale of

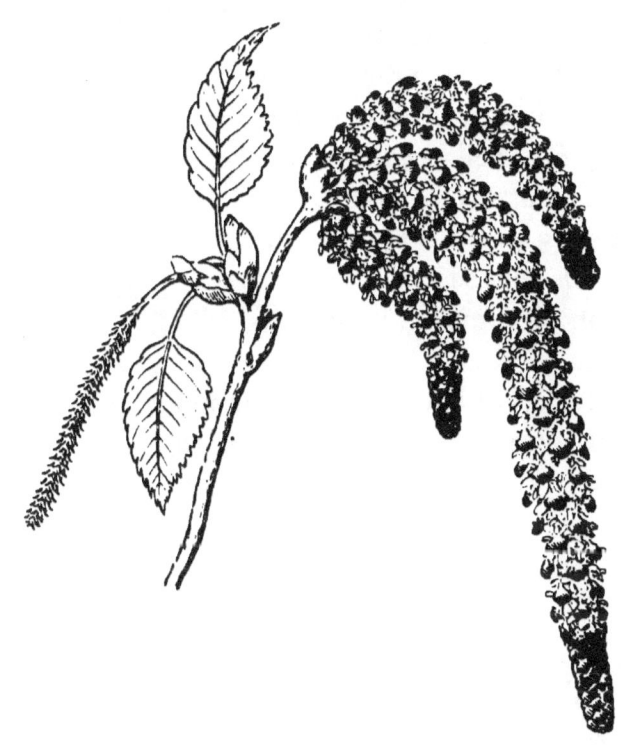

FIG. 22.—Flowering branch of Birch (*Betula alba*), with catkins and, one cone.

the female cone has two bracts and three flowers. There is no calyx in either male or female flower.

In the catkins of the Hazel the pair of bracts become soldered to the scale. Each scale bears four

[1] See page 279.

stamens, which are deeply cleft and apparently double. The cone of the Hazel consists of a number of overlapping scales, in the midst of which can be discovered by careful search several bracts, each of which acts as a sheath to two flowers. Every female flower bears two carpels, which are indicated by the two crimson styles, but in ripening one cell becomes suppressed, and each ripe ovary encloses, as a rule, only a single seed. Double-seeded filberts are not, however, very uncommon.

FIG. 23.—Scale of Hazel-catkin, with male flowers. Magnified. See also p. 290.

The flowers of Hazel first appear, in their unexpanded state, in autumn; the catkins pass the winter without external protection, but the female flowers are wrapped up within the enveloping scales. In March the styles lengthen, pushing their way through the apex of the cone, where their crimson colour makes them very conspicuous. Only a few of the female flowers persist; two, three or four may set their seeds and produce nuts. The bracts grow steadily through the summer, and form envelopes around the nuts, and these envelopes have the form and often the colour of leaves. The same thing may be observed in the

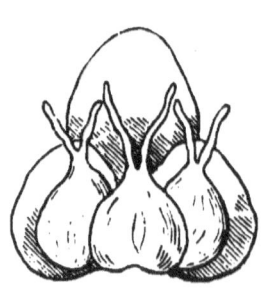

FIG. 24.—Three pistillate flowers of Birch, with their enclosing scales. Magnified.

G

Hornbean, where the bracts seem adapted to aid in the dispersal of the fruit. This cannot be the case with the heavy nuts of the Hazel, yet the structure is there, though applied to some new use, of which I can give no account. The cupule of the Acorn corresponds to the bracts of the Hazel and Hornbean.

I will next describe the flowers of the Willow, the "palms" of country people, so called, I believe,

FIG. 25.—Young cone of Hazel, after flowering. The flowers are grouped in pairs and enclosed by bracts, which enlarge after flowering

because they are plentiful in most years on Palm Sunday.

Alder, Birch and Hazel bear both catkins and cones on the same branches, Willow bears them on different trees. Alder, Birch and Hazel are called *monœcious*, Willow *diœcious*. Both kinds are included under the term *diclinous*. I must rebel against these ugly words, so ill-chosen (though the great Linnæus is responsible for the first two) that even when you know their etymology, you can hardly understand or remember

CATKINS 83

them any the better. Let us try whether we cannot for our immediate use find something less objectionable. *Unisexual* (of one sex) is plainer than *diclinous;*

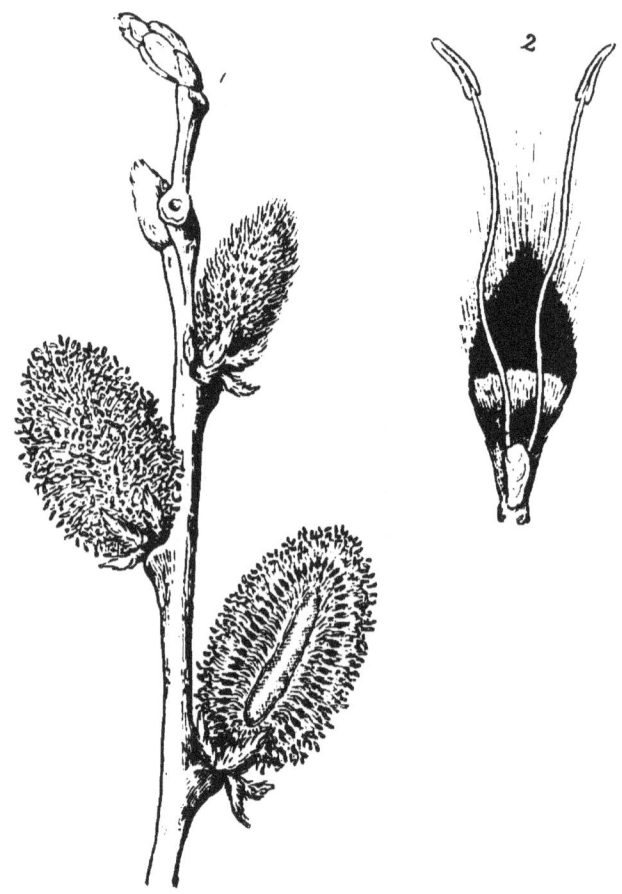

Fig. 26.—1. Catkins of Willow, entire and in longitudinal section; 2. Staminate flower of ditto. Magnified.

we may perhaps be allowed to substitute *incompletely unisexual* for *monœcious*, and *completely unisexual* for *diœcious*.

Willow trees, as I have said, are wholly male or wholly female, *completely unisexual;* so are Poplars. Find a male Willow of the Common Sallow kind, and watch its flowers open from day to day. The scales of the bud part, and a mass of silvery hairs shows itself, which lengthens, turns yellow, and at length seems to be made up of stamens and pollen. If you break such a catkin across, you will find it made up of a vast multitude of silky bracts, each of which bears two stamens. There is a minute gland to each bract, which exudes sweet juice, and helps us to understand why the Sallow is attractive to Insects.

The female flowers are borne many together on spikes (we cannot call them cones, though they answer to the cones of Alder, Birch and Hazel). Each flower is ensheathed by a bract, and contains a seed vessel or ovary mounted on a stalk, and ending above in a forked style. The ovary contains many seeds.

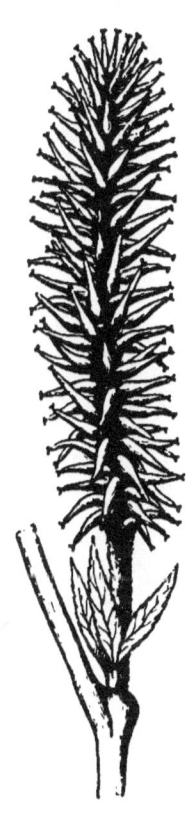

FIG. 27.—Cone of pistillate flowers of Willow.

In June, look out for the ripe pods of the female Willow. The halves of the ovary separate at the top, and gradually curl themselves in opposite directions, exposing to view a multitude of silky seeds. I have found it very amusing to pull out a tuft of Willow seeds, and put them in the sun. They seem as if they were alive, pushing one another away, and slowly expanding into a great fluffy mass, which is easily wafted to a distance by a light current of air. The spreading out of the pinch of hairy seeds is due to the fact that each seed bears a crown of hairs,

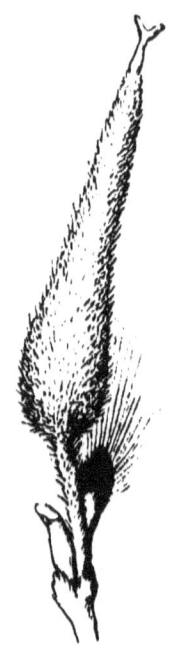 

FIG. 28. — Pistillate flower of Willow. Magnified.

FIG. 29.—Ripe fruit of Willow, bursting. Magnified.

which lie close while the seeds are in the pod, but stand out like rays or spokes as they dry. It is easy to see the advantage to the Willow of getting its seeds spread out into a ball so light in proportion to its surface that any wind of summer can blow it far over the fields. This is a common contrivance, but almost every fresh case has its own peculiar features. The Poplar, the Bullrush and the Willow-herb (so named from the shape of the leaves) are curiously like the Willow in the mode of dispersal of the seeds.

The catkin is a form of inflorescence particularly well-suited to wind-fertilisation. The pollen-grains are formed in vast numbers within the crowded flowers of the long catkins; they are easily shaken out of these dangling tassels, wafted by the wind to great distances, and lodged on the branching styles of the female cones. The Willow gets help from Insects, especially Bees, which visit its nectar-bearing flowers on bright days. Willows will set their seeds though there is no male plant within a mile.

Why do catkins appear so early, before the leaves are out? It may be that the leaves would interfere with the dispersal of the pollen by wind. They would inevitably catch much of the pollen wafted to or from the tree, so that there is a distinct advantage in getting the flowers fertilised before the leaves appear.

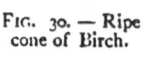

Fig. 30. — Ripe cone of Birch.

The pollen of Willow does not depend upon wafting by the wind, yet the Willow flowers before it comes into leaf. Yes, and it secures two considerable advantages thereby. Its catkins are far more conspicuous on bare boughs and the Bees are very glad to visit its honey-bearing flowers in early spring, when flowers are so few.

How can we explain the different lengths of the cones or spikes of female flowers? The Willow produces very many female flowers, as also does the Birch; Alder notably fewer; Hazel often about eight only, of which less than half usually ripen. The difference becomes still more conspicuous if we compare the number of the seeds produced. The Willow may produce a million (I have not counted them) on a single spike; Birch fewer, but still very many; Alder perhaps a tenth of the number of the Birch; Hazel only two, three or four. The size of the seeds is naturally in an inverse proportion to the number, and the quantity of food stored up in the seeds will vary almost directly with the size. What is the meaning of these striking differences? Willow, Poplar and Birch seeds are dispersed by the wind; therefore they must be light, and since many will be lost, it is desirable that they should be very numerous. I think that the seeds of Alder are dispersed by running water, for they spring up along the banks of rivers and brooks.

FIG. 31.—Ripe cone of Birch, broken across to show the winged fruits. Magnified.

Seeds so dispersed need not be very light, and it is probable that a smaller proportion is wasted than in wind-dispersed seeds. The Hazel-nut is eatable, and is carried off by Squirrels or other animals. Here and there one is dropped, or buried and forgotten, and this is able to germinate. A small seed would tempt no animal, but a large seed protected by a hard shell is worth carrying off, and yet has a chance of germinating after all.

Why are the pods of the Willow soft and green, while the cones of the Alder and Birch and the nuts of the Hazel are woody? Probably because the minute seeds of the Willow ripen quickly and are easily dispersed. They require no protection against the rains and frost of winter, as the slow-maturing seeds of Alder, Birch and Hazel do.

One question more. Why are trees so often completely unisexual? Where many flowers are borne upon one plant, as is commonly the case with trees, they would infallibly fertilise one another continually, if all were perfect. By the complete separation of the stamens and pistils, self-fertilisation becomes impossible.

Annual plants are hardly ever completely unisexual. The transport of the pollen from one plant to another, whether by Insects or by the wind, is an operation which might conceivably be hindered in a particular year by deficiency of a particular species of Insect, by perfectly still weather, or by long-continued rain. Such accidents, even though they came round but once a century, or once in a thousand years, would greatly reduce the numbers of an annual plant, and might even exterminate it. But it would signify little to a tree that the whole crop of seeds should fail in a particular year.[1]

[1] See Darwin's *Cross- and Self-fertilisation of Plants*, Chapter X.

There are other questions about catkins which I do not propose to my readers because I have found them so far insoluble. Insoluble questions are plenty as blackberries; the art of the investigator of nature is to put questions which have some chance of getting answered.

The Oil-beetle (Meloe).

A few days ago (April 2) I came across a female Oil-beetle (Meloe) walking on a bank in the sun, and seeking for a place in which to deposit the multitude of eggs which distended her enormous abdomen. She at last chose a grassy place and began to dig. When she had made a hole large enough to contain her body, she turned round and pushed her abdomen into the hole. Here she remained quite still for a long time, her head being just visible. I marked the place, came home, and spent an hour or more in reading Newport's history of the Oil-beetle.[1]

When I came back the Meloe was gone and the hole was closed with earth. It was easy to dig out the cluster of eggs, which were very numerous (Newport says 3,000-4,000 in the first laying), very minute, of an orange-yellow colour, adhering together and lying all one way. Following Newport's example, I placed the eggs in a box filled with earth, and awaited the appearance of the larvæ, which emerged on May 8th (five weeks). Newport's observations on a captive Meloe teach us that the abdomen re-fills with

[1] *Linn. Trans.*, Vol. XX. pp. 297-357, Pl. XIV. (1851); Vol. XXI. pp. 167-183, Pl. XX. (1853).

eggs after laying, and that four packets may be deposited in three weeks. After each deposition the body shrinks and the Insect feeds voraciously.

When the period of hatching is completed the eggs burst, and countless larvæ, minute, six-legged, and of bright yellow colour, emerge. They possess con-

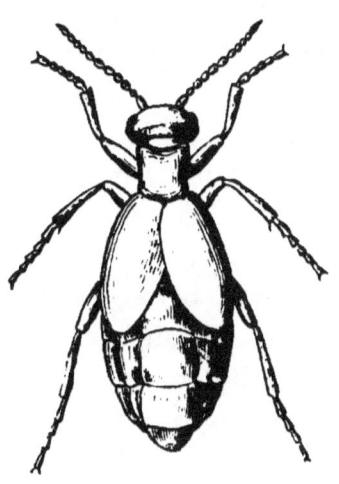

FIG. 32.—Female Oil-beetle (*Meloe proscarabeus*). Natural size.

spicuous black eyes, pointed mandibles, and legs furnished at the extremity with a claw and two lateral hooks, so that they appear three-parted. The whole larva is about $\frac{1}{12}$ in. long. After a time it runs and creeps with great agility by means of its hooked legs. It can also crawl upon a smooth, vertical surface, such as an upright sheet of glass, by attaching alternately the front legs and a pair of false feet, which project from the last segment, and resemble the claspers of a caterpillar. The body is moved either forwards or backwards in a looping fashion.

THE OIL-BEETLE (MELOE)

It is easy enough to go so far, and to obtain small active larvæ from the eggs of Meloe. We should at first be inclined to suspect that these larvæ would seek their food in the ground, grow big, pupate, and change to perfect Beetles. But this does not happen. If the larvæ are imprisoned, even though they may be supplied with food of any imaginable description, they soon perish. If they are kept warm and exposed to light, they run about busily, seeking to escape; if they are kept in the dark, they remain still. In either case, they perish in two or three weeks.

It had been noticed before Newport's time that such larvæ were occasionally to be found clustered on grass, or on the flowers of Buttercup and Dandelion. The same or very similar larvæ had been found clinging to the hairs of various Bees and Flies, and Latreille had thought it probable that they attached themselves to flower-visiting Bees, and were conveyed to the nest, which at this time of the year is being stored with pollen and honey.

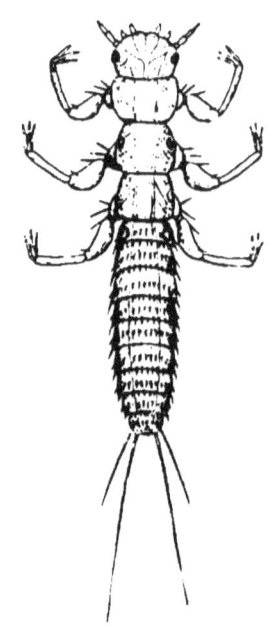

FIG. 33.—First larva of Oil-beetle, magnified. After Newport

Newport placed a burrowing Bee in a bottle which contained a brood of these larvæ. They instantly seized any part of the Bee which came within reach, leg, wing, or hairs, and mounted in crowds upon the body, causing the greatest uneasiness to the Bee, which struggled in vain to detach them. Newport remarks that all the Hymenoptera on which Meloe larvæ have been found burrow in the ground, and all the Flies on which they have been taken are, like Volucella, parasitic in the nests of such Hymenoptera. The statement is too sweeping, for Meloe larvæ have been known to attach themselves in mistake or despair to other flying Insects, as will appear later. Newport himself gives us one instance. Having secured three or four hundred Meloe larvæ in a bottle, he put with them several living Weevils and a single Malachius (a small flower-haunting Beetle). The Weevils were not disturbed, but the larvæ instantly attached themselves to the Malachius in such numbers as to cover it and deprive it of the power of moving; most of them clung on for many hours.

Newport found pupæ and advanced larvæ of one species of Meloe among the nests of a particular burrowing Bee (Anthophora), and now the case seemed to be pretty complete. It remained to see the young larva brought to the nest, and to watch its operations on arrival. He took with him in June fresh-hatched larvæ of *Meloe proscarabæus* (the commonest species) and *M. violaceus*, and placed them in the nests of *Anthophora retusa*, each cell of which then contained a Bee-maggot and a supply of pollen-paste, its proper food. At first he thought that the experiment was succeeding, for one of the larvæ seemed to attack the Bee-maggot with its mandibles. But nothing further happened. He left the intruders in the nest and went away. Next day

THE OIL-BEETLE (MELOE)

the Bee-larvæ with their stores of food were still there, but the Meloe-larvæ had gone.

Thus baffled, Newport called to mind that it was *Meloe cicatricosus* which he had found in Anthophora nests, and that he had experimented with the larvæ of two other species. He was, it appears, unable to repeat the experiment with the larvæ of *cicatricosus*, or to discover the true hosts of *violaceus* and *proscarabæus*. But he knew perfectly well the next stage of *Meloe cicatricosus*. This he had found in the closed cells of Anthophora in considerable numbers. It is a curved, cylindrical, thick, almost footless grub, motionless and of pale orange colour. The head is small, and the hinder end of the body encumbered with the cast skin of the active larva. He supposed that the active larvæ devours the egg of its host, changes its skin, and then feeds upon the honey. In August, by which time it had greatly increased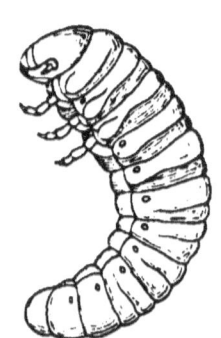

FIG. 34.—Second larva of Oil-beetle (*Meloe cicatricosus*) magnified. After Fabre.

in size (Newport found advanced larvæ ¾ in. long), it changes to a pupa, and the perfect Beetle soon emerges. This hibernates in the same cell, and only emerges in the following spring. The newly emerged Beetles are small and shrunken. But they feed greedily upon the leaves and flowers of the Buttercup or some few other spring flowers, and become plump. The Beetles are fond of sunshine. They drink water plentifully, and in captivity require their food to be frequently wetted. The females are much larger than the males. Newport tells us that the males are exceedingly pugnacious, and often fight, depriving each other of one of the antennæ. The unwieldy form of the female, the inky purple colour, and the sluggish gait have something unpleasant about them, and few would care to handle the insect. Those who do so find that it sweats from every joint drops of yellow fluid, probably acrid. As if aware of its ill-

FIG. 35.—Pseudo-chrysalis (inactive larva) and pupa of Oil-beetle (Meloe). After Newport.

taste, the Beetle creeps leisurely about in full sunlight, as fearless and as conspicuous as a Skunk. What a variety of experiences it has gone through! It has lived three lives, each requiring its own instincts and a distinct bodily structure, when its last packet of eggs is laid and the curious history of greed and cunning comes to an end.

Such is the account which Newport was able to give. We shall see that it does not by any means exhaust this complicated life-history. The fresh information we owe to Fabre.[1] I am glad of the opportunity to introduce his graphic sketches to some English readers who may not have met with them before. The translation is somewhat condensed.

Meloe cicatricosus, he remarks, infests the nests of two other Anthophoræ besides the *A. retusa* spoken of by Newport, but though our author had found the larvæ in the cells of its hosts, he had never seen the female wandering in search of a place to lay her eggs. The history of his investigations dates from May 23rd, 1858, and the scene was a steep slope bordering the road from Carpentras to Bédoin. The slope, baked in the sun, was the abode of swarms of burrowing Bees (Anthophora). A scanty turf extended from the edge of the road to the foot of the slope. To observe the Bees more closely, Fabre lay upon the grass, when his clothes became covered with what looked like little yellow fleas, rushing about with desperate haste. These Insects, which gave him the appearance of being dusted with ochre, were soon recognised as the active larvæ of Meloe.

On the patch of grass were some few flowers, and among these a Groundsel and a Chamomile. Observing these closely, Fabre saw that almost all the flowers were occupied by the larvæ. They might be found crouching motionless among the florets of a single head. None were to be seen on a Poppy or a Crucifer which grew by the side of the Composites.

"Besides the hordes encamped on the heads of the Composites," continues Fabre, "which were motion-

[1] *Nouveaux Souvenirs Entomologiques*, XVI. (1882).

less, as if they had accomplished their immediate purpose, I could see a still more numerous crowd whose restless movements showed that they were still unsatisfied. On the ground among the grass ran about in disorder innumerable little larvæ, resembling the swarms which issue from an overturned ant-hill. Some mounted in haste a blade of grass, and came down again with equal speed; others clung to the cottony surface of dry Cudweed, but after resting there a moment, quitted it and resumed their quest. In a space some ten metres square, there was hardly a blade of grass which had not been explored by the larvæ.

"I had plainly before me a swarm just hatched from a mass of eggs laid in the earth. Some of the larvæ were already settled among the florets of Chamomile and Groundsel, waiting the arrival of Bees, but most of them had still to discover a suitable station. The whole multitude, which must have amounted to many thousands, could hardly be the offspring of one female, in spite of all that Newport says of the astonishing fertility of this Insect.

"The strip of turf stretched a long way by the side of the road, but no Meloe larvæ could be seen except within a few square metres adjacent to the burrows of the Bees. The larvæ could not have travelled far; there were no laggards, such as inevitably follow a marching column. It would appear that the Meloe, not laying her eggs at hazard, or leaving the young to make their own way to the burrows which they were destined to inhabit, chooses a place haunted by Anthophora, and lays her eggs there.

"So great was the multitude of Meloe larvæ that sooner or later almost all the Bees of the neighbourhood must have become infested with them. Though comparatively few larvæ had yet gained the flowerheads, and though the Anthophora seldom alights on the ground, where the Meloe larvæ were most plentiful, almost all the Bees which I caught and examined had several larvæ entangled in the hairs of the thorax.

"I have also found Meloe larvæ on Melectes and Cœlioxys, two Hymenoptera which are parasitic upon Anthophora. Quitting for a moment their bold dartings to and fro in front of the galleries of the Bee, still in process of construction, these thieves of honey stored up for others settle for an instant upon a flower of Chamomile, and then it is the robber who is robbed. A tiny larva creeps upon their downy covering. When the parasite, having destroyed the egg of the Anthophora, comes to lay its own egg on the honey, the Meloe larva lets itself down, destroys the second egg, and remains sole proprietor of the food within the cell. The store of honey laid up by the Anthophora will thus belong to three owners in turn, and will remain with the weakest of the three.

"But who can say that the Meloe may not itself be dispossessed by some new thief, or even, while still a soft, fat and sluggish larva, fall a prey to some destroyer, who will devour its living entrails? As we ponder over the deadly and remorseless strife which nature prescribes to these various creatures, by turns wealthy and destitute, devourers and devoured, a feeling of horror mingles with the admiration excited by their artifices. We forget that these things pass in a lower sphere of existence, and shudder at the train of thefts, deceptions and robberies which enter into the plan of our *alma parens rerum*.

"The young Meloe larvæ, once settled on the hairs of Anthophora or its nest-parasites, are in a sure way to reach sooner or later the cell which is their goal. Are they guided by intelligence, or do they attach themselves by chance to anything that offers? It was not difficult to clear up the point. Drone-flies and Blow-flies dashed now and then against the flowers of the Groundsel and Chamomile in which the Meloe larvæ were lurking, and paused for a moment to suck the sweet juices. On all, or very nearly all these Flies, I found Meloe larvæ, clinging to the thoracic hairs. An Ammophila (Sand-wasp) which provisions its burrows in early spring with a caterpillar, just grazed the surface of a flower. I caught it, and found Meloe larvæ running over its body. Neither the Eristalis and Blow-fly, whose larvæ feed on putrid flesh, nor the Sand-wasp which feeds its young with caterpillars, could ever transport the Meloe larvæ to the cells filled with honey which they desire. These larvæ had gone astray, and it is clear that their instinct is not infallible.

"Let us study more carefully the Meloe larvæ as they wait in expectation on the Chamomile flowers. There they are, ten, twelve or more together, half concealed within or between the florets. They are not easily seen, as the amber-yellow of their bodies matches the colour of the central florets. So long as the flower is undisturbed, they remain motionless, head downwards. One might suppose that they were sucking nectar, but if that were their aim, they would move about from one floret to another, which does not happen, except when they seek the most advantageous position after a false alarm. The florets are a mere lurking-place, and they will eat nothing until their jaws crush the egg of an Anthophora.

"Let us gently explore a Chamomile flower with a straw. The larvæ quit their retreats, and run out along the white ray-flowers. When they gain the extreme tip, they attach themselves by the appendages of the tail, or perhaps by means of a viscid secretion, such as the Sitaris larvæ possess. With the body extended in free air and the legs free, they sway about in all directions, as if bent upon reaching some object out of easy reach. If nothing approaches which they can grasp, they by and by return to the centre of the flower, and become motionless once more.

"If we bring any object within their reach, they attach themselves with extraordinary quickness. A blade of grass, a straw, the arm of a forceps—anything will do, so eager are they to quit their retreat. But they soon find out when they have made a mistake, and run up and down, trying to get back to the flower, if it is still possible. After having once grasped a straw, they will not, if replaced on the flower, be entrapped a second time so easily.

"I tried little bits of cloth or velvet, torn off my clothes, as resembling more or less the hairy covering of a Hymenopterous Insect, plugs of cotton-wool, and flocky pellets stripped off the Cudweed. Any of these the larvæ clutched at, but instead of remaining close and still, as they do on the Insect, they perceived that they were out of the right track, and tried to escape. This might have been predicted, for I had seen them running restlessly to and fro on the woolly Cudweed. If a mere hairy surface would satisfy them, almost all the larvæ which fell in with these plants would perish there without an effort.

"I next tried bringing a live Anthophora to the flower. The Bee, cleared of any parasites which it happened to bear, was held by the wings, and made to touch the flower for a moment, when it was invariably found to bear Meloe larvæ, which clung to its hairs. The larvæ climb instantly upon the thorax, and there they remain motionless; the second stage of their migration is now accomplished.

"In the same way I tried all the live Insects which I could immediately procure—Drone-fly, Blow-fly, Honey-bee and small Butterflies. All were instantly overrun by the Meloe larvæ. What is more, no subsequent attempts were made to regain the flower. There were no Beetles at hand, so I did not make trial of them. Newport, working under different conditions, for his larvæ were imprisoned in a bottle, while mine were under natural conditions, saw them climb upon a Malachius Beetle, and remain there at rest, from which I conclude that I should have got the same result with Beetles as with Flies (Eristalis, for instance). I have since found a large Beetle, the Rose Beetle, which continually haunts flowers, beset with Meloe larvæ. When all the Insects I could procure had been tried, I offered them a large black Spider. They climbed upon it without hesitation, reached the articulation of the legs, and remained there motionless. Anything, it appears, suffices to tempt them from their temporary retreat; they attach themselves to the first living thing which comes in their way. Hence the necessity of a vast number of eggs; the great majority of the larvæ go astray, and never gain the cells of the Anthophora. Fertility supplies the defects of instinct.

"How does the Meloe larva quit the Bee, which has guided it to the cell? With larvæ found upon various Hymenoptera I made some trials, such as Newport had previously made. Meloe larvæ when brought near to the larvæ and pupæ of Anthophora, paid no attention to them; others placed close to cells filled with honey, did not enter them and at most touched the edge; while such as were placed inside the cells, came out immediately or perished by suffocation.

"Dug-out nests of Anthophora had previously informed me that *Meloe cicatricosus* is parasitic upon this Bee, in whose cells I had found the adult Meloe, dead and dry. The yellow larvæ, found alive upon Anthophora, had been recognised by Newport as the larvæ of Meloe. Bearing in mind these facts, which impressed me the more as I had recently been investigating the similar history of Sitaris, I betook myself on May 21st of the following year to Carpentras, and visited the nests which the Anthophora was now engaged in constructing. I felt pretty sure of finding out sooner or later the life-history of Sitaris, which was very plentiful, but was less hopeful about the Meloe, which, though it occurs in the same nests, is very rare. After six hours of digging I secured many cells enclosing Sitaris, and two with Meloe. On the dark and liquid honey was a wrinkled membrane, and upon this a yellow larva. The membrane was the empty envelope of an Anthophora egg; the larva was the larva of Meloe.

"The Meloe larva quits the hairs of the Bee just when she lays her egg. Since contact with honey would be fatal, the tactics of the Sitaris are pursued, and the larva drops upon the egg as it is laid. The next step is to devour the contents of the floating egg, and after this meal, the only one which it takes in its active stage, it undergoes a kind of transformation, feeding hereafter upon the honey stored up by the Anthophora. Hence the obstacle which rendered fruitless my own previous attempts as well as those of Newport. It is useless to offer to Meloe larvæ honey, larvæ of Anthophora, or pupæ; they must attain the freshly deposited egg."

Fabre has described in detail the subsequent transformations of the Sitaris larva, of which the English reader will find an interesting account with illustrative figures, in Lubbock's *Origin and Metamorphoses of Insects* (Chap. II.). The history of Meloe is very similar. The active larva changes to a soft grub, which feeds exclusively on honey. A third larval form, not unlike a Lamellicorn larva, succeeds. About Midsummer this is transformed into what Fabre calls the pseudo-chrysalis. The Meloe in this stage, still encumbered by the cast skin of the third larva, had been observed by Newport, who remarks that the stout mandibles and hooked feet of the third larva seem suited for digging, perhaps for making a passage to a fresh cell, whose honey is unexhausted.

The pseudo-chrysalis is sluggish, with a yellow, horny skin, and from it the true pupa at length issues. The only perfect Beetle which Fabre reared came out in September, but the Insect does not show itself nor seek its mate till the following spring.

THE CORN-RIGS OF BEAMSLEY FELL.

April 15, 1895.—I look out from my study-windows across Wharfedale, and see faint but unmistakable tokens of the advance of spring. The level meadows in the floor of the valley are green, and not grey as they were a month ago. The woods, which all winter through were of uniform black, are now resolved into masses of feathery trees, softly pencilled with brown and green. A grove of poplars in Denton Park is distinguished by the colour and the branching at a distance of a full mile.

When we come to close quarters we see the green buds of the Sycamore, the brown buds of the Poplar and the russet flowers of the Elm. The trees which slept are waking.

It is near sun-down, and the sun shines straight down this reach of the valley, bringing out with his horizontal rays the faintest surface-markings. On the opposite hill-side the drainage furrows are ruled in close and regular lines across the meadows and pastures. I can also distinguish much slighter furrows here and there, which are corn-rigs, the furrows of ploughed land. These last are so ill-defined that I should not have recognised them, if I had not lately walked past those very fields. Falling into talk with a farm-labourer, he pointed out which pastures had been reaped within his own memory. Higher up the valley are terraces, once cultivated by the spade, and there are many signs that corn and vegetables formerly flourished where now all is grass. Such traces of ancient tillage are not peculiar to Wharfedale or to Yorkshire. They are common in all parts of England. Canon Raine points out that in parts of Tynedale which have never been tilled within living memory the Black Book of Hexham Priory shows that corn was once raised. "If this evidence were wanting, the lay-riggs, as they are called, which the rich sward of untold years has been unable to obliterate, still show decisively that in days long gone by the ploughman and the sower have been there."[1] Marshall in 1804 found that over all the country from the Tamar to the eastern border of Dorsetshire, open commons which had never been ploughed within the memory of man were marked with ridge and furrow.[2]

Corn-rigs on grass land do not necessarily prove a diminution in the acreage of tilled land. In our day the few ploughed fields of this part of Wharfedale belong to the lower and richer lands. But when drainage was rare and insufficient, these low-lying fields were considered too wet and too liable to inun-

[1] *History of Hexham Priory*, Vol. II. Preface, p. xviii. (Surtees Soc.).
[2] Quoted by Prothero, *Pioneers and Progress of English Farming*.

dation, and the uplands were preferred for raising corn. I have noticed in old maps of Yorkshire towns that flat meadows near rivers were often occupied by tenter-hooks for stretching cloth upon. Probably they were too wet for crops. But the higher ground, up to 600 ft. or more above sea-level, often bore crops. Wheat would not thrive at the greater elevations, but oats and especially rye would do well enough. Rye can be harvested late, in cold and wet weather, and this was no doubt the chief reason why it was so largely grown in the north of England down to the early part of the eighteenth century.

There are few published records which give information respecting the history of agriculture in Wharfedale. Craven was long a wild and backward district. In Edward III.'s time[1] the labourers of Craven with the inhabitants of other desolate regions, were made exempt from the prohibition to wander in search of work in summer time. In Henry VII.'s reign there was a notable increase of population in Yorkshire, as the many parish churches of that age testify, and this increase would encourage the tillage of lands previously waste. This was also a time when much arable land was laid down in grass, as we learn from the statute-book and from the complaints of Hugh Latimer and Sir Thomas More. Whether the acreage of tilled land positively declined in the sixteenth century it would be hard to say. But there is no doubt that during the next three hundred years there was a vast increase in the ploughed land of all parts of England, an increase which went on steadily

[1] 25 Edward III. Stat. 2, Cap. 2.

until the middle of the nineteenth century. During the last fifty years, however, there has been a marked concentration of population in Yorkshire. Outside the great towns there has been a considerable aggregate increase, due to the growth of manufacturing villages, health resorts, and suburban houses, all dependent upon manufactures. But the proper rural population has declined. Unroofed cottages are common in Wharfedale, and a far larger number have been swept away altogether. There was a time when princes and parliaments would have interposed to check the evil by such blundering enactments as they could devise. We have learnt by experience that mankind cannot be effectually driven to adopt an occupation and a place of abode which accord with the views, not of themselves but of their rulers.

The importation of cheap corn has materially reduced the area of ploughed land within recent years. In every part of England are great expanses of pasture and meadow which waved with corn within the memory of men still living. The day is at hand when the vast majority of Englishmen will dwell in cities. A hundred years ago the vast majority were rustics.

Was it better to live in England then than now? Better in some ways, no doubt. Men worked, if they did not sleep, in pure air, and in sight of the trees and the blue sky. The simple out-of-door pastimes of Shakespeare's day were better than the music-hall and the street-organ. There were then no wildernesses of streets to cut the children off from the very possibility of knowing the face of Nature.

Some things have remained much the same through

centuries of change. Curiously enough, among the things which abide with us is political and social restlessness. Almost every age has dimly felt that it was on the eve of startling changes. Such changes have come and passed, and left the old fabric standing. Wat Tyler and John Ball in the fourteenth century were followed by Jack Cade in the fifteenth, by the pilgrimage of Grace and Robert Kett in the sixteenth, by the Levellers and Clubmen in the seventeenth, and by the Luddites and Chartists in the nineteenth. The hopeful and the ignorant and the clamorous are always full of "a good time coming," which will never come, it is to be feared, until men learn self-denial.

Some things have changed for the better during three centuries. Among these are food, lodging, clothing, education and medical treatment. People feed much better and live much longer than they did. If they are not a good deal wiser, it is their own fault.

Take one thing with another, I would rather live in Yorkshire in 1895 than in 1495, 1595, or 1695. Between 1795 and 1895 it would be harder to choose. The changes of the last hundred years are very conspicuous, but they are very hard to equate.

The Cuckoo.

April 19.—This morning I heard for the first time this year the "two-fold shout" of the Cuckoo. One male Bird is certainly here. I see him fly to and fro across the fields, as if seeking a mate.

April 22.—We have now several Cuckoos in our valley, the males as yet greatly predominating. The males are more fixed in their abode than the females, which rove a good deal and pick up several mates.

May 20.—A Cuckoo's egg found in a Wagtail's nest. The small Birds are certainly afraid of the Cuckoo, who pursues them as if to see where they are going to lay. The Hawk-like appearance no doubt adds to the terror which the Cuckoo inspires.

We have few memoirs on the habits of Birds more interesting than Dr. Jenner's *Observations on the Natural History of the Cuckoo*, published in the *Philosophical Transactions* for 1788.

Jenner, the discoverer of vaccination, was a man of varied tastes and acquirements. He was accomplished in music and studied Natural History with diligence and success. In this subject he had the advantage of instruction by a first-rate master, John Hunter, in whose house he lived for two years, and to whom he addressed, for communication to the Royal Society, his memoirs on the Cuckoo. In 1788 Jenner was thirty-nine years old, and practising medicine at Berkeley. During the same year he came up to London, in order to make known his views as to the relation between cow-pox and small-pox, which were coldly received by the great physicians. His first case of successful vaccination was still some years in the future (1796).

At one time it had seemed likely that Jenner might become a professed naturalist. He had been employed, probably on Hunter's recommendation, to

THE CUCKOO

arrange the zoological collections brought back by Sir Joseph Banks from Captain Cook's first voyage of discovery, and the post of naturalist on the second voyage had been offered to him. But love of his Gloucestershire home and the prospect of distinction in medicine decided Jenner to remain in England. Throughout his eminent and useful career Natural History continued to occupy his attention. He studied the hibernation of the Hedgehog, the fossils of Gloucestershire, and the habits of Birds His long-meditated paper on the Migration of Birds was sent in to the Royal Society in 1823, the year of his death.

Jenner's Observations on the Cuckoo are too long for verbatim quotation, but as they are not accessible to every reader, I think it may be worth while to give them in a condensed form. I preserve, as far as possible, the words of the author.

"The first appearance of Cuckoos is about the 17th of April. Like other migrating Birds they arrive and depart in succession, and are more numerous in the second than the first week of their arrival. The song of the male soon proclaims its arrival. The song of the female is widely different, and I believe that few are acquainted with it; the cry of the Dab-chick bears the nearest resemblance to it.[1]

"Unlike most Birds, Cuckoos do not pair. The female does not begin to lay till some weeks after her

[1] Some naturalists are of opinion that the female bird calls "Cuckoo" like the male; but clear and direct testimony cannot be quoted. See Newton in Yarrell's *British Birds*, 4th edition, Vol. 2, and *Zoologist*, June, 1886.

arrival. I never could procure an egg till after the middle of May.[1]

"The Cuckoo makes choice of the nests of a great variety of small Birds. I have known its egg intrusted to the care of the Hedge-sparrow, the Water-wagtail, the Titlark, the Yellow-hammer, the Green-Linnet, and the Whinchat;[2] among these it generally selects the three former, but shows a much greater partiality to the Hedge-sparrow than to any of the rest. The Hedge-sparrow commonly takes up four or five days in laying her eggs. During this time (generally after she has laid one or two) the Cuckoo contrives to deposit her egg among the rest. When the Hedge-sparrow (or other Bird) has sat her usual time, and disengaged the young Cuckoo and some of her own offspring from the shell, her own young ones and any of her eggs that remain unhatched, are soon turned out, the young Cuckoo, which is commonly hatched first, remaining possessor of the nest, and sole object of her future care. The young Birds are not previously killed, nor are the eggs demolished; but all are left to perish together, either entangled about the bush which contains the nest, or lying on the ground under it.

"*June* 18, 1787, I examined the nest of a Hedge-

[1] The Cuckoo is said by Dr. Rey to lay an egg every other day (sometimes every day for a short time) from the middle of May to the middle of July.

[2] To this list we may add the Wren, the Red-backed Shrike, the Bunting, and the Redstart. There is a fuller list in Harting's *Summer Migrants*, pp. 222-3. No fewer than 110 species of birds are recorded as having been known to hatch the eggs of the Cuckoo.

THE CUCKOO

sparrow which then contained a Cuckoo's and three Hedge-sparrow's eggs. On inspecting it the day following I found that the nest now contained only a young Cuckoo and one young Hedge-sparrow. The nest was placed so near the extremity of a hedge, that I could distinctly see what was going forward in it, and to my astonishment, I saw the young Cuckoo in the act of turning out the young Hedge-sparrow. The mode of accomplishing this was very curious. The Cuckoo, with the assistance of its rump and wings, contrived to get the other Bird upon its back, and making a lodgment for the burden by elevating its elbows clambered backward with it up the side of the nest till it reached the top, where, resting for a moment it threw off its load with a jerk, and quite disengaged it from the nest. It remained for a short time feeling about with the extremities of its wings, as if to be convinced that the business was properly executed, and then dropped into the nest again. I have often seen it examine, as it were, with the extremities of its wings, an egg and nestling before it began its operations, and the nice sensibility which these parts seem to possess, compensated the want of sight, which as yet it was destitute of. I afterwards put in an egg, and this by a similar process, was conveyed to the edge of the nest, and thrown out. These experiments I have repeated several times in different nests, and have always found the young Cuckoo disposed to act in the same manner. In climbing up the nest, it sometimes drops its burden, and thus is foiled in its endeavours, but after a little respite, the work is resumed, and goes on almost incessantly till it is effected. This disposition for turning out its companions declines from the time it is two or three days old. I have frequently seen a young Cuckoo, hatched nine or ten days, remove a nestling that had been placed in the nest with it, but suffer an egg, put there at the same time to remain unmolested. The singularity of its shape is well adapted to these purposes, for, different from other newly hatched Birds, its back from the scapulæ downwards is very broad, with a considerable depression in the middle. This depression seems formed by nature to give a more secure lodgment to the egg or the young Bird, when the Cuckoo is employed in removing either of them from the nest. When it is about twelve days old, the cavity is quite filled up, and then the back assumes the shape of nestling birds in general.

"*July* 9.—A young Cuckoo, that had been hatched by a Hedge-sparrow about four hours, was confined in the nest in such a manner that it could not possibly turn out the young Hedge-sparrows which were hatched at the same time, though it was almost incessantly making attempts to effect it. The consequence was, the old Birds fed the whole alike, and appeared in every respect to pay the same attention to their own young as to the young Cuckoo, until the 13th, when the nest was unfortunately plundered.

"The smallness of the Cuckoo's egg in proportion to the size of the Bird is a circumstance that hitherto, I believe, has escaped the notice of the ornithologist. So great is the disproportion, that the egg is in general smaller than that of the House-sparrow, whereas the difference in the size of the Birds is nearly as five to one. Eggs produced at different times by the same Cuckoo vary very much in size. I have found one that weighed only forty-three grains, and another that weighed fifty-five grains. The colour is extremely variable: some, both in ground and pencilling, very much resemble the House-sparrow's; some are indistinctly covered with bran-coloured spots; and others are marked with lines of black, resembling in some measure the eggs of the Yellow-hammer.

"The circumstance of the young Cuckoo's being destined by nature to throw out the young Hedge-sparrows, seems to account for the parent-Cuckoo's dropping her egg in the nest of Birds so small as those I have particularised. If she were to do this in the nest of a Bird which produced a large egg, and consequently a large nestling, the young Cuckoo would probably be unable to throw out the young Birds. I have known a case in which a Hedge-sparrow sat upon a Cuckoo's egg and one of her own. Her own egg was hatched five days before the Cuckoo's, and the young Hedge-sparrow gained such a superiority in size that the Cuckoo was unable to lift it out of the nest till the Cuckoo was two days old.

"It appears a little extraordinary that two Cuckoos' eggs should ever be deposited in the same nest, as the young one produced from one of them must inevitably perish; yet I have known two instances of this kind, one of which I shall relate.

"*June* 27, 1787.—Two Cuckoos and a Hedge-sparrow were hatched in the same nest this morning; one Hedge-sparrow's egg remained unhatched. A few hours after a contest began between the Cuckoos, which continued undetermined till the next afternoon, when one of them, which was somewhat superior in size, turned out the other, together with the young Hedge-sparrow and the unhatched egg. The contest was very remarkable. The combatants alternately appeared to have the advantage, as each carried the other several times nearly to the top of the nest, and then sunk down again, oppressed by the weight of its burden, till at length the strongest prevailed, and was afterwards brought up by the Hedge-sparrows.

"Why should not the Cuckoo, like other Birds, build a nest, incubate its eggs, and rear its own young? There is no reason to be assigned from the formation of the Bird, why it should not perform all these several offices. May not the singularities of the Cuckoo be owing to the short residence this Bird makes in the country where it propagates, and the call of nature to produce during that short residence a numerous progeny? The Cuckoo's first appearance here is about the middle of April, commonly on the 17th.[1] Its egg is not ready for incubation till some weeks after its arrival, seldom before the middle of May. A fortnight is taken up by the sitting Bird in hatching the egg. The young Bird generally continues about three weeks in the nest before it flies, and the foster-parents feed it more than five weeks after this period; so that [even] if a Cuckoo should be ready with an egg much sooner than the time pointed out, not a single nestling would be fit to provide for itself before its parent would be instinctively directed to

[1] In other parts of England the Cuckoo often arrives a few days earlier.

seek a new residence : for old Cuckoos take their leave of this country the first week in July.

"The Cuckoo goes on laying till the eve of her departure from this country, for though old Cuckoos in general take their leave the first week in July, I have known an egg to be hatched in the nest of a Hedge-sparrow so late as the 15th.

"Among the many peculiarities of the young Cuckoo there is one that shows itself very early. Long before it leaves the nest it frequently, when irritated, assumes the manner of a Bird of prey, looks ferocious, throws itself back, and pecks at anything presented to it with great vehemence, often at the same time making a chuckling noise, like a young Hawk. Sometimes, when disturbed in a smaller degree, it makes a kind of hissing noise, accompanied with a heaving motion of the whole body. The growth of the young Cuckoo is uncommonly rapid. The chirp is plaintive, like that of the Hedge-sparrow; but the sound is not acquired from the foster-parent, as it is the same whether it be reared by the Hedge-sparrow or any other Bird. It never acquires the adult note during its stay in this country.

"The stomachs of young Cuckoos contain a great variety of food, animal or vegetable. Hedge-sparrows in general feed the young Cuckoo with scarcely anything but animal food; the Titlark feeds it principally with grasshoppers. In one fed by Hedge-sparrows, the contents of the stomach were almost entirely vegetable, such as wheat, small vetches, etc. This served to clear up a point which before had somewhat puzzled me; for having found the Cuckoo's egg in the nest of a Green-Linnet, which begins very early to feed its young with vegetable food, I was apprehensive till I saw this fact that this Bird would have been an unfit foster-parent for the young Cuckoo.

"There seems to be no precise time fixed for the departure of young Cuckoos. I believe they go off in succession, probably as soon as they are capable of taking care of themselves. Though they stay here till they are nearly equal in size and growth of plumage to the old Cuckoo, yet the fostering care of the Hedge-sparrow is not withdrawn from them. I have frequently seen the young Cuckoo of such a size that the Hedge-sparrow has perched on its back or half-expanded wing, in order to put the food into its mouth. At this advanced stage, I believe that young Cuckoos procure some food for themselves. If they did not go off in succession, it is probable that we should see them in large numbers by the middle of August, but they are not more numerous at any season than the parent-birds in May and June."

The habits of the Cuckoo afford a tempting field for speculation, and many attempts have been made to trace the probable origin of the instinct which leads this Bird to lay her eggs in the nests of others. Jenner, as we have seen, looks upon the necessity of early migration from the north as the determining cause. But the early migration is still an unexplained fact. Is it an antecedent or a consequent? Does the Cuckoo lay her eggs in other Birds' nests, in order that she may leave early, or does she leave early because she has no young brood to detain her in the north? Has she a motive independent of her young for retreating in July? We cannot tell.

It has been supposed that the parasitic egg-laying of the Cuckoo depends upon the circumstance that the eggs instead of being laid daily, mature in succession, with intervals of two or three days. If all the eggs were hatched in the same nest, the operation would be protracted, and inconvenience would result from the existence of eggs, young nestlings and older nestlings in the same nest. This actually happens in the case of the American Cuckoo which is nonparasitic. Such a negative exception is not a refutation, but we have no proof that the rate of formation of the eggs is a fixed and unalterable condition, capable of dictating the mode of incubation.

It is interesting to note that the habits of the Cuckoo are not absolutely determined by obvious facts of structure, and also that other Birds exhibit the beginnings of a possible parasitic instinct. Some species of Cuckoo build their own nests, hatch their own eggs, and feed their own young. The common American Cuckoo is one of these. Our common Cuckoo has been said to lay her eggs on the bare ground, to hatch them, and to feed the young.[1] There are several truly parasitic Cuckoos besides our familiar species. One of these is European, three are Australian.[2] This affords the possibility of deciding

[1] Mr. Harting suggests that there may be an error of observation here, and that the Nightjar has been taken for a Cuckoo.
[2] Ramsay, quoted in Darwin's *Origin of Species*, Chap. VII.

which facts of structure and life-history are and which are not necessary to the parasitic mode of incubation. It would seem probable that the small size of the egg is directly connected with parasitism. Whether early migration or even migration at all is an essential condition I do not know. By comparison of various species of parasitic Birds it appears that they are prone to lay their eggs in the nests of Birds whose eggs are somewhat similar in size and colour. This tendency perhaps exists in our common Cuckoo, though the contrast between her eggs and those of the Hedge-sparrow is notorious. There is some reason for supposing that the colour of the eggs laid by every female is peculiar and constant. Each Cuckoo returns, it is believed, year after year to the same place, and lays her eggs in the nests of one particular species only. Dr. Rey supposes that each Cuckoo keeps to the nests of that species by which she was herself reared.[1] Certain Birds which are not Cuckoos at all regularly lay their eggs in the nests of other Birds. Among these are more than one species of Icteridæ, some of which are named Orioles, though not belonging to the family of true Orioles. Various Birds of the most diverse kinds have been known to practise the same trick casually. The Starling's egg for instance, has been found in a Woodpecker's nest. Much work remains to be done in the way of collecting, authenticating, and com-

---

[1] *Alles und Neues aus dem Haushalte des Kuckucks* An interesting discussion of the question is to be found in Harting's *Summer Migrants*, pp. 224-8.

paring information before an adequate history of the instinct can be related.

Our Cuckoo is said to be unable from its size to sit upon the nests in which its eggs are commonly laid, and therefore unable to lay in the usual fashion. Several witnesses, apparently trustworthy, are quoted as having seen the Cuckoo carry her egg in her bill. One observer watched a Cuckoo through a telescope, saw her lay her egg on a bank, and then carry it in her bill to a Wagtail's nest.[1] Dr. Rey quotes a case of a Cuckoo's egg smeared with red earth similar to that which covered the ground about the nest.

It has been repeatedly said that it is the female Cuckoo or the Birds to whom the nest belongs, which turn out the nestlings.[2] Jenner's narrative, which is very explicit, has however been confirmed by subsequent observers, and appears to be entitled to full credit. Montagu saw a young Cuckoo repeatedly throw out a young Swallow put into the nest for the purpose of experiment. Blackwall saw a nestling Cuckoo turn both young Birds and eggs out of the nest in which he had placed them for the purpose. Mrs. Blackburn made a clever drawing of a young

---

[1] This and other cases are given in Newton's *Dictionary of Birds*, which contains much curious information respecting the habits of the Cuckoo.

[2] It is strongly maintained by X. Raspail in a recent paper (*Mém. Soc. Zool. de France*, 1895) that the hen Cuckoo watches the process of hatching, and as soon as the chicks begin to free themselves, destroys the eggs with her beak. She throws the eggs or the young Birds out of the nest as soon as her own egg is hatched.

Cuckoo in the act of ejecting a nestling Pipit. (See Gould's *Introduction*, or Harting's *Summer Migrants*). John Hancock saw a young Cuckoo make repeated and at length successful efforts to throw out the eggs and nestlings of a Hedge Sparrow. (*Nat. Hist. Trans. Northumberland and Durham*, Vol. VIII.; reprinted in *Zoologist*, May, 1886.) The accounts of Montagu, Blackwall, and Mrs. Blackburn are fully related in Harting's *Summer Migrants*.

Like some other Birds, the Cuckoo changes his note after the breeding season. The cry becomes hoarser, the first syllable is sometimes doubled, and the musical interval between the two sounds is altered.

Jenner only slightly refers to one singular feature of the Cuckoo, viz., its resemblance to a Sparrowhawk. Many inexperienced people have been deceived by it. The barred plumage of the chest, belly and legs are the chief means of deception, but there is also a resemblance of attitude. The small Birds seem to be imposed upon, for they show terror at the sight of a Cuckoo, desert their nests and build new ones when intruded upon by her, or at other times collect and chase her as they would chase a Sparrowhawk. It is common to see a Cuckoo followed like a Hawk by a small bird, and late in summer a young Cuckoo is not unfrequently mistaken for a Hawk by some man or boy, and shot. It seems likely that the Hawk-like appearance of the Cuckoo intimidates the sitting Bird, and causes her to offer less resistance to the invasion of her nest. But we have still much to learn about the difficulties and artifices of the Cuckoo.

BUDS.

April 20.—The trees are fast coming into leaf. It is a good time for observing the structure of buds, and seeing how they expand. Let us begin by

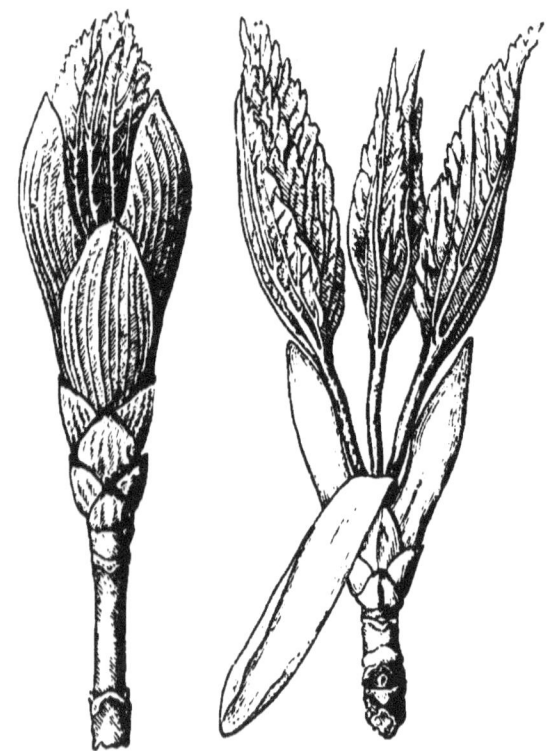

FIG. 36.--Bud of Sycamore, expanding, with outer scales and inner folded foliage-leaves.

examining a Sycamore-bud, which is big and of simple structure.

I gather a Sycamore-bud which is bursting, and the first thing which catches the eye is that it is enveloped in a number of tough scales. The outer scales and the tips of the inner ones have been long exposed to the air, and are dark-coloured; the parts which were concealed in the unexpanded bud are paler. There are four rows of scales, two opposite rows of four each, and two intermediate rows of three each; there are therefore fourteen scales to the bud, or sometimes twelve only. All these are carried upon a short stem. If we strip off all the scales, one by one, we shall find two pairs of folded foliage-leaves in the centre of the bud. This is most easily seen in a bud which has already expanded, and whose parts are enlarged.

On the ground beneath the Sycamore hundreds of bud-scales are lying about. They are deciduous, and are cast as soon as their purpose has been served. We may conclude from this that they serve only for the protection of the folded leaves within. Pick up a few of the fallen bud-scales. They are rather long and narrow, deeply concave on the side which faced inwards, and well shaped for wrapping round the bud. At the top of each scale is a small knob or point. Look at it with a lens. You will see (more distinctly in some than in others) a small three-lobed projection. This is often curved round to the inner side of the scale, and sometimes hidden by a mass of brown hairs. On some of the larger scales, which were next to the folded leaves, the projection at the summit is five-lobed and quite leaflike. We can hardly be wrong if we call it a rudimentary leaf-blade. In the Flowering Currant, as well as in some other trees and shrubs, this leaf-blade often attains a fair size, turns green, and remains for a long time attached to the branch by its supporting scale. If the tip of the scale is a rudimentary leaf, what is the scale itself? A leaf-stalk, surely. That seems a natural and almost

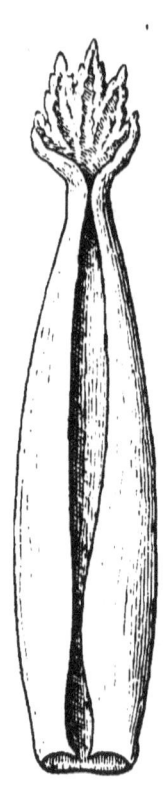

FIG. 37.—Bud-scale of Sycamore, with rudimentary leaf. Magnified.

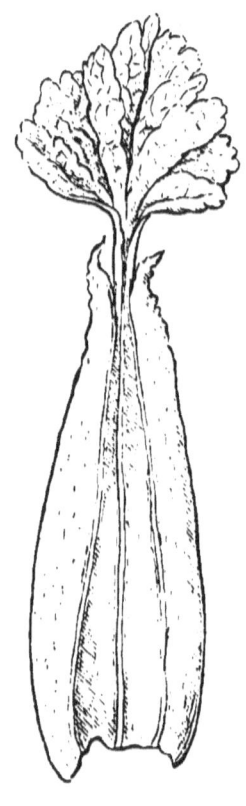

FIG. 38.—Bud-scale of Flowering Currant, with rudimentary leaf. Magnified.

inevitable answer, but it is wrong, as we shall see by and by.

Some Sycamore-buds contain bunches of flowers as well as leaves. These are larger, and expand earlier than those which produce leaves only.

Let us now turn to the true foliage-leaves in the centre of the bud. They are folded up fan-wise. Why is this?

A few weeks ago it was hard frost, but the Sycamore-buds were already fully formed and exposed to the air. They were to be seen all through the winter,

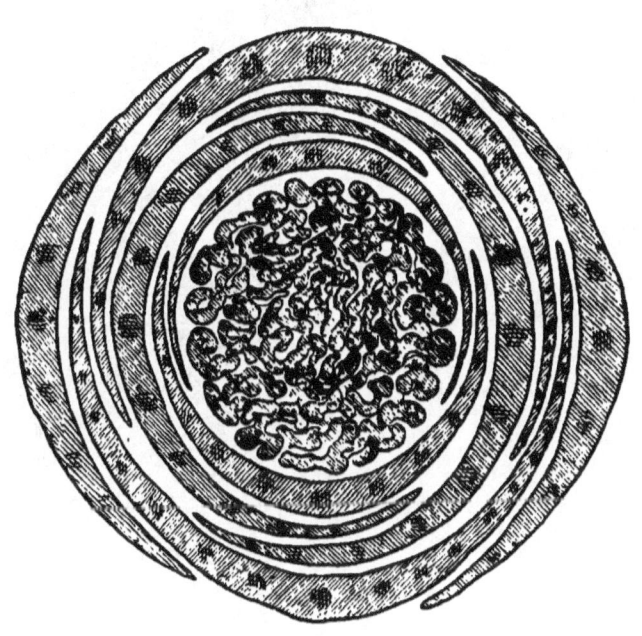

FIG. 39.—Cross-section of Sycamore, showing scales and folded foliage-leaves. Magnified.

and had to endure all the cold and wet of the severe season. One obvious precaution is to restrict as much as possible the exposed surface of the leaves. Hence leaves in the bud are packed up tightly, sometimes folded, sometimes crumpled up, sometimes rolled round. The hardships of winter explain why the leaves are enveloped in scales. The scales keep off cold air and moisture.

We can imitate the folding of the Sycamore-leaf by a paper model. Take a sheet of paper, cut it to a roundish shape, and fold it along the middle. On

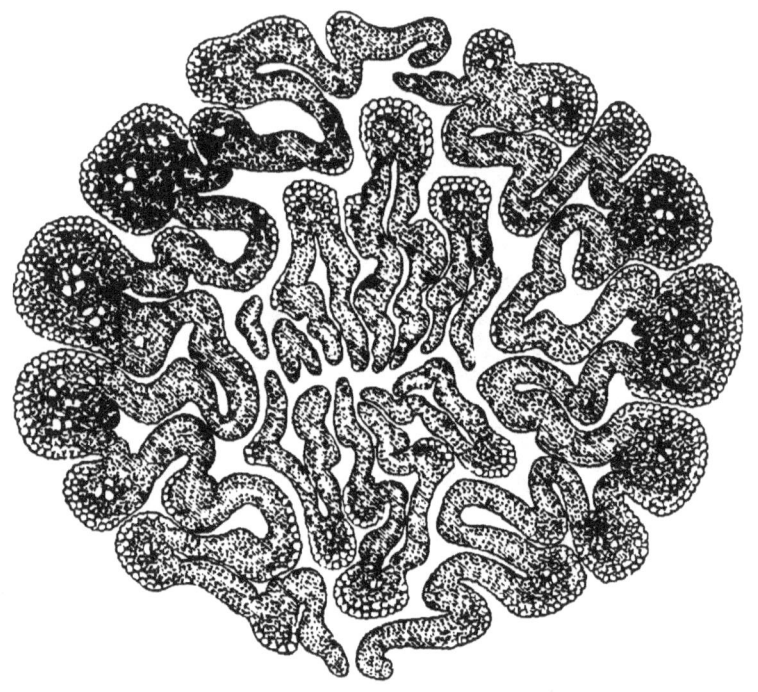

FIG. 40.—Cross-section of folded foliage-leaves from bud of Sycamore (two pairs). Highly magnified.

each side of the first fold make two symmetrical folds. To imitate the Sycamore-leaf as closely as possible the spaces between the folds must be narrow towards the base, and widen out towards the tip, as in a fan. When folded up tight, the paper model will have a pointed base and a square end. Such a shape is not good for close packing. The square ends of four leaves would give the bud a great bulging tip. Let us bring our paper model to a point by cutting obliquely through the folds near the apex. Observe that the cut must not interfere with the midrib, lest the leaf be weakened, but should slope towards the opposite side of the folded leaf. When the paper is

FIG. 41.—Leaf of Sycamore.

unfolded after being cut through, it will be five-fingered. Four such leaves, pointed at base and apex, will go into a neat oval bud. The five-fingered Sycamore-leaf is well-shaped for packing, but I will not say that this is its only merit.

We will take the bud of the Beech as our next example. It is long, slender and pointed. In spring it swells and lengthens; the numerous bud-scales part, and several brown membranes (stipules) appear between them. As the bud expands more fully, green leaves push out from among the brown stipules. As in the Sycamore the leaf is folded. From the midrib about ten pairs of lateral ribs are given off, and the thin, green blade is sharply folded between each pair. The ribs and the margin of the leaf are fringed with silky hairs, which entangle much air, and so screen the delicate young leaf from cold winds or fierce sunlight. In Horse Chestnut, the great White Willow, and some other trees the leaves are downy when they first appear, but cast all their hairs before long. The shoot enclosed within the bud of the Beech grows fast, and the stipules soon become widely spaced, then the leaves and leaf-stalks are fully seen, and we observe that the stipules spring in pairs from the bases of the leaves. Each stipule is a long curly, strap-like blade, which withers and falls off as soon as the leaf is fully expanded.

Stipules do not always fall off early. In Hawthorn, Lady's Mantle, Pansies, and many other plants they form small green leaves of peculiar shape, which last all summer. What are stipules? They are lateral outgrowths from the leaf-base, which develop early, and enclose the central part, or leaf proper. They are often a protection to the unexpanded leaf, and where this is their sole function, they are deciduous.

The principal leaf-blade and the stipules, if there are any, spring from a particular part of the leaf, which we have called the *leaf-base*. This is flattish, and of inconsiderable length in the full-grown leaf. If a leaf-stalk appears at all, it is of later formation than the base and blade, and appears between them. In the rudimentary leaf of the bud-scale of Sycamore no leaf-stalk forms because the development of the leaf is checked in an early stage. Hence the bud-scale itself is not properly, in the Sycamore, a leaf-stalk, but a greatly enlarged leaf-base.

By close examination we can satisfy ourselves that the bud-scales of the Beech are not leaf-bases, as in the Sycamore, but stipules. Between each pair there is a minute green leaf, which never develops. The bud-scales are the stipules of several such undeveloped leaves.[1]

The bud of Lilac exhibits some interesting peculiarities. Here the branch does not usually end in a single bud, as in most trees, but the terminal bud fails to develop, and a pair of lateral buds take its place; hence the strong tendency to fork which we find in the branches of Lilac. The central leaves are enveloped in four rows of bud-scales, alternately two or three in a row. So far there is nothing out of the common. But if you dissect away the bud-scales and examine them one by one, you will find that they pass by insensible gradations into ordinary foliage-leaves. The outermost scales are triangular, the next longer and narrower at the base, and so on. There are no rudiments of blades at the tips. The bud-scales are not here enlarged leaf-bases, but small leaves, the blade being more and more developed as we pass inwards.

Bud-scales are not therefore all formed exactly in

[1] Goebel, *Bot. Zeitung*, p. 774 (1880).

the same way. Some buds have no scales at all. Thus tropical plants, if they have no dry season to face, and some evergreens (Ivy, Box), bear naked buds.

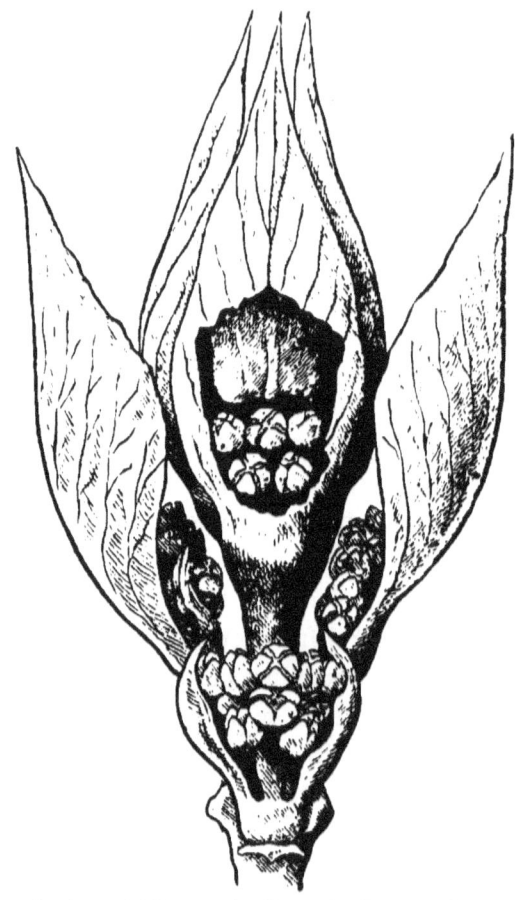

FIG. 42. - Flowering bud of Lilac, partly dissected. Some of the scales have been removed, and part of the central leaf has been cut away to show the flowers within. Magnified.

Some Conifers have scales (Pine, Spruce, Yew); others none (Cypress, Juniper). If there are bud-scales, they may be simply leaves of small size and simple

K

form (Lilac, Honeysuckle). This is rarely the case with plants which possess divided leaves. Such plants have bud-scales which are enlarged leaf-bases, rudiments of the blade and of the stipules being often visible at the tip. Sycamore, Horse Chestnut, Ash, and most Rosaceous trees are examples. The true foliage-leaves may also be enveloped by stipules, which may either be few, green and leaf-like, as in the Alder and Poplar, or numerous, brown, and purely protective. Oak and Beech buds are of this latter kind.

A bud is a new shoot, complete or nearly so in all its parts, but as yet unexpanded. It consists of a stem with leaves and leaf-like appendages, and perhaps flowers too. A bud is a thing of slow growth, for though it may only be a quarter of an inch long, it contains a number of perfect, if minute leaves, or flowers, or both. When spring comes round every sunny day is of value, and no time must be lost. Everything is therefore made ready beforehand. The leaves and flowers are all there, of microscopic size and crowded into the smallest space, but with the encouragement of moderate warmth, they soon swell out and unfold.

The buds of trees and shrubs and other perennial plants are formed months before they expand. Look at a tree in summer or autumn. Close to the base of each leaf, in the angle between it and the branch, or in some cases on the scar of a leaf which has fallen off, you will find small buds. These small buds are those of the following year. It is by no means uncommon to find by careful search at the side of the bud of the following year, still younger buds, and when required these can be pushed forward rapidly. Sometimes a late frost kills the newly-expanded leaves, say, of a Beech. This happened in the year 1891. A hard frost on the night of Whit-Sunday killed nearly all the young and tender leaves, and for weeks after all the Beeches looked brown and withered. But before midsummer the buds, which in the usual course would have expanded in 1892, had already pushed forth, and each of these showed at its base a bud which had been hastened a year, and which at length expanded in the spring of 1892 instead of 1893.

Drought, or the devastations of Insects may bring about the same results. Sometimes mere luxuriance of growth accelerates the development of the buds, and what would in the regular course form winter-buds expand and develop into shoots in July. Such fresh summer shoots are common in Horse Chestnut. Elder, and Sycamore for instance.

We can produce these results at pleasure by removing the leaves from young shoots in spring. When the buds are thus made to expand a year in advance, the leaves whose development had been arrested, in order that they might be converted into bud-scales, resume their growth, and expand into fully formed foliage-leaves or transitional forms connecting these with bud-scales. It is as if the tree perceived that leaves and not bud-scales would be wanted immediately. New buds form in the axils of the leaves thus hurried on, and it is these which open in the following spring.

The protection of leaf-buds is effected in various ways. The bud-scales are sometimes downy, as in buds of the Willow; sometimes they pour forth a sticky substance, made of resin or gum, especially at the time when the bud is just ready to open. Such secretions are found in the buds of the Horse-Chestnut and the Black Poplar. Hairs protect the bud from both cold and wet, chiefly by enclosing air, which cannot easily be dislodged from very narrow spaces. Imprisoned air is a very bad conductor of heat, and it does not allow water to penetrate. Even when the bud-scales are not downy, the thin layers of air between them are a great protection. Sometimes the bud-scales are excavated by broad and very thin air-spaces. Many buds, especially of herbaceous plants, are buried beneath the ground.

But for disturbing circumstances, which are, however, inevitable in the case of trees and shrubs, every leaf would develop a bud in its axil, that is, in the angle between it and the main stem. It is easy to see that the shoots would become terribly crowded if every leaf produced its bud, and every bud formed a shoot. But poorly illuminated leaves often produce no buds in their axils, or the buds fail to expand. At the tips of the branches, on the other hand, where the light is profuse, large buds, developing strong shoots, will appear. Hence the branching is most vigorous in an upward and outward direction. Many variations are to be observed, and these lead to variations in the mode of branching, and therefore in the form of the full-grown tree. If the terminal bud is larger than any of the lateral ones, and expands regularly upwards, we shall get a spiry tree. In an Oak there is a terminal bud, around which several lateral ones are clustered; this arrangement gives the rosette-like branching which we all know. In Lilac, as we have already seen, the terminal bud always fails to develop, and a pair of lateral ones take its place; hence the strong tendency to fork which we observe in this shrub. In the Elm and Lime also the terminal bud fails to develop, though it is often vigorous up to a certain point. Here the lateral bud next below takes up the running, and pushes out very nearly in the line of a regularly expanding terminal bud.

There is no constant position for the flower-buds of trees and shrubs. They may be terminal, but are more commonly lateral, as in the Willow. Sometimes there is no separate winter flower-bud at all, as in Beech and Oak. Here the flowers appear in the axils of an ordinary leafy shoot. The flower-buds are often, however, of special size and shape, and enclose a leafy or leafless inflorescence. They may often be distinguished from ordinary winter-buds weeks before they expand, and the flowers can be made out by opening or cutting across the bud as early as the previous autumn or summer.

Many plants make use of their buds as means of propagation. All the organs necessary to a plant are present in a bud or can be readily formed upon it. Stem and leaves are already there. Roots can be pushed out from the stem when required. The leafy stem can form flowers when flowers are wanted. If the bud is to grow into an independent plant, it must in general be detached from the parent. See how this is managed in the Strawberry or the creeping

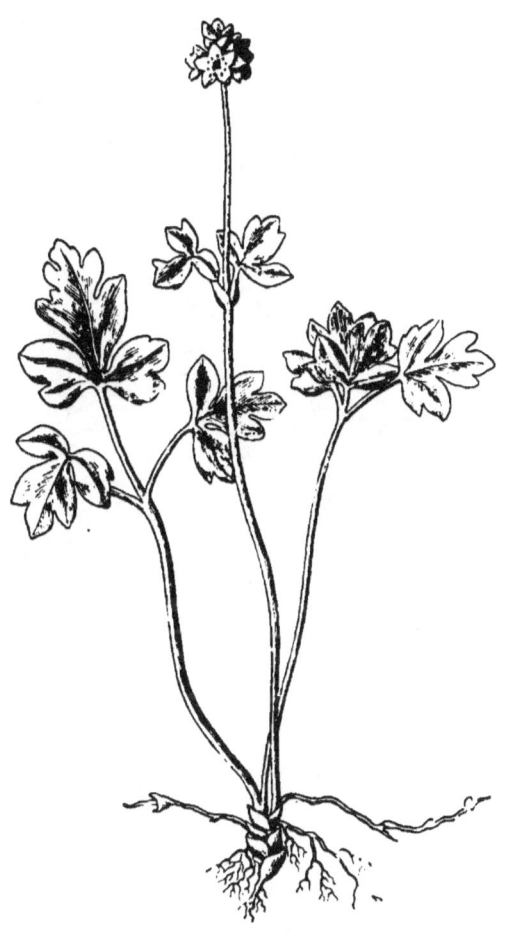

FIG. 43.—Moschatel. (*Adoxa moschatellina.*)

Buttercup. They push out, sometimes very rapidly, a stalk or runner, which lengthens and makes its way over the ground. The runner bears one or more buds, which are thus carried to a place where they have room to establish themselves. Before long they become rooted, and send up new upright stems. The Celandine forms numbers of little green buds which break loose and are scattered (how, I do not know). You may sometimes find them in hundreds and thousands, lying loose on a lawn. These little buds are capable of growing into full-sized plants. The

FIG. 44.—Head of Flowers of Moschatel.

Ladies' Smock and the Water-cress form little bulbs on the leaves in the same way.

Flower-buds often exhibit beautiful arrangements for close-packing. It is hard to see without a microscope the very minute flowers, and the way in which they are arranged long before the bud opens. But I will mention one case where the need for close-packing seems to govern the shape and arrangement of the fully-opened flower.

In some parts of the country the little Muscatel is

a common hedge-row flower in April or May. It bears five, small, greenish flowers on one flower-stalk. Many young naturalists have gathered these clustered cups, and have wondered to see that the side-flowers are differently made from the single flower at the top.

FIG. 45.—Flowers of Moschatel, *a*, uppermost flower; *b*, ditto, seen from beneath; *c*, lateral flower, seen from beneath.

Each side-flower is five-pointed, bearing five petals, ten stamens, and five carpels. But the top flower is four-pointed, and has only four petals, eight stamens, and four carpels. Why this difference? We will suppose that the number of the flowers in the head is determined by causes not known to us, and that five has proved to be the most convenient number. We will also take it for granted that the arrangement of the five flowers into a compact head is beneficial to the Muscatel, though the reason is not known to me at least. Then the necessity for neat packing in the bud requires that the side-flowers shall be five-pointed, and the top flower four-pointed. The whole head is of nearly globular shape, with six faces regularly placed all round. Each flower occupies one face, the sixth and bottom one being wanted for the flower-stalk. Take an apple, and cut it square by paring off the sides. We shall get four vertical faces, which indicate the places of the side-flowers. They will not be exactly square, but each will have two upright straight edges, a semi-circular edge above, and a flatter, curved edge below, near to the stalk. Such a face is nearly pentagonal or five-cornered, and a five-cornered flower will do very well for each of the four side-places. But the top-flower must fit in between the four side-flowers, and to do this neatly it must be four-pointed. A five-pointed flower could only fit very awkwardly into the squarish place at the top of the flower-head.

If the flowers of Muscatel were not so crowded together they might be all alike. So too, where a great many small flowers are packed together into one bud, they may all have the same number of points, and this we find in the flower-buds of umbel-bearing flowers, such as Cow-parsnip. In the flower-bud of Cow-parsnip there is a dense crowd of unexpanded flowers, all five-pointed. But where the head consists of only five flowers, four beneath and one on the top, it cannot, so far as I know, be neatly and closely packed in any other way than that which we see in the Muscatel.

THE BOTANY OF A RAILWAY STATION.

May 10.—The platform of the little station where we get in and out of the train every day was well asphalted five or six years ago. The pavement is still sound and good, except in a few places near the palings, where plants have pushed beneath it, heaved it up, and at length made their way to the air and light. Shoots of Coltsfoot and the common Field Equisetum, sent out from plants well established on the adjoining slope, have succeeded in breaking through a solid stratum more than an inch thick.

I do not in the least understand how the growing shoots of herbaceous plants can force their way through an asphalt pavement. It is true that any observant person can find like instances. We know of the sapling which grew through the hole in the middle of a millstone, and ended by lifting the millstone from the ground. We have perhaps seen, as I myself have done, a tree growing in the cleft of a rock, and at length forcing asunder fragments which would tax the strength of several men to lift. We talk of turgidity and the like, but we have not solved the problem. Where is the mechanician who will undertake to push a growing herbaceous stem, neither so thick nor so firm as a lead pencil, through an inch of hard asphalt?

On the same platform are little hollows, hardly apparent to the eye, where the asphalt has been chipped or indented. In some of these hollows dust has collected, and seeds or spores have germinated there. A little annual grass, a chickweed and a moss flourish in these minute garden-plots, few of which are bigger than a sixpence. At the foot of one of the lamp-posts a handful of earth and sand has collected, and here five sorts of flowering plants have managed to establish themselves.

Germs of living things are scattered everywhere, and some develop in the most unexpected situations.

Let us take one group of plants—the Fungi, and one group of animals—the Insects. I cannot find room for more than one or two examples of each. Fungi have been known to thrive in the saturated solution of copper sulphate used in a Daniell's cell. Various species find nourishment in almost every animal or vegetable tissue, alive or dead, raw or manufactured. Insects are known to feed upon organic matter of every kind. Glacier ice harbours one species in countless numbers. A small Beetle feeds upon argol (crude potassium tartrate), and has lived and propagated for years in a stoppered bottle half full of this substance, which is kept in my laboratory.

We speak of life as a precious thing, and such it really is. But we must admit that it is not precious because of its rarity. There is an unlimited supply of life of all kinds ; it is food and opportunity which run short. Malthus and the new Poor Law have interpreted nature truly enough. Population of every kind is always tending to outrun the means of subsistence. Of course it cannot actually do so, or cannot do it long. There is consolation for the anxious observer of Man and nature in the very obvious reflection that intolerable evils, such as starvation, are deadly, and work their own cure. The fear of starvation is an evil too, but it is one of those evils which brings forth good. All human arts and activities orginate in the fear of starvation, and decline when it is removed. The natural contrivances which delight us by their ingenuity and completeness are just as much the outcome of difficulties about food and space as human arts.

Over-population may be a source of many evils, but it is the mainspring of Life.

SUMMER TWILIGHT.

June 21.—This evening we were able to play bowls till nearly ten o'clock. It is true that the last shots were made with some difficulty. It was necessary to hold a white handkerchief or even a lighted match over the jack to show where it was. But the finish was exciting, and we persisted till five minutes to ten. At that time the north-western sky was still bright, and we smoked on the terrace for a long time, watching the slow fading of the light and the gradation of the colour from amber to purple and grey. Some one began to talk about the long twilight of midsummer, and mentioned the well-known fact that it lasts till dawn. Then an argument arose about the length of twilight in different seasons of the year. All were agreed that it lasts longest at midsummer, but opinions differed as to whether it is shortest at midwinter or at the equinoxes. I have since tried to inform myself a little on this point.

The duration of twilight can only be stated precisely when a somewhat arbitrary assumption is made. Twilight is light received only by reflection from particles, solid or liquid, which float in the air. Hence it varies according to the state of the sky. In high alpine regions there may be no twilight, although the sun is only a little way below the horizon. But under favourable atmospheric conditions twilight lasts till the sun has sunk 18° below the horizon. How long after sunset will it be before the sun sinks so low, and how will the season of the year affect the time?

Draw a circle to represent the apparent path of the sun in the heavens, and draw chords in the circle to represent the horizon on particular days. In summer the chord will be low down, and the diurnal path will be greater than the nocturnal. In winter the chord will be high up, and the nocturnal path greater than the diurnal.

Parallel to each chord, and beneath it, draw a line, to represent the limit of twilight. The arc intercepted between the two lines will measure the duration of twilight. It will be seen that the arc intercepted will be shorter, the nearer the two lines approach the equator of the circle, shortest of all when one is as much above the equator as the other is below it. That position will occur twice in the year, a little before the vernal equinox, and a little after the autumnal equinox. At the equinox itself the horizon will coincide with the equator of the circle, and the limit of twilight will be 18° below it. As we approach the solstices the intercepted arc will increase in length, but as the limit of twilight is always below the horizon, it will come nearer to the equator at the winter than at the summer solstice. The maximum for the year will be at midsummer. In London twilight never ends between May 22 and July 21. There is a second maximum in midwinter, when twilight lasts about 2 hrs. 10 min. The minima fall about February 28 and October 12, when twilight lasts only about 1 hr. 50 min.

Midsummer Blooms.

One of the glories of the summer is the abundance of white flowers, not merely scattered about the lawns and hedges like stars, but clustered into sheets and masses. Hawthorn, Elder, Meadow-sweet, the great Umbellifers, Apple, Pear, Bird-cherry, Mountain Ash, Horse Chestnut and Guelder Rose are familiar instances. The spectacle opens in May, and ends in August with the great Wood Campanulas. As the autumnal equinox draws near, the twilight is too short and the nocturnal Insects too few for flowers of this particular kind.

It seems probable on a first consideration of the question that these expanses of white flowers, glimmering in the twilight of the short night of Midsummer, are lures to night-flying Insects. Some of them offer fragrance as well as contrast of colour, and the fragrance is often more powerful by night. But when we come to note what Insects have been actually seen to visit the great white blooms, we shall find that some of them are visited only in the daytime, and by various kinds of Flies. The Umbelliferæ are adapted for fertilisation by Flies, and their odours, often disagreeable and rank to our taste, seem to be well suited to the appetite of Flies. The night-haunted blooms on the other hand are largely visited by Moths, and belong more particularly to the season when Moths are plentiful. Moths, it would seem, enjoy the same odours as ourselves, for many of the perfumes which attract Moths delight mankind also.

Midsummer is not however reserved for any one kind of Insect, or any one kind of flower. It is the very height of the flowering season, when the profusion, though not the variety, of flowers is greatest. Then all sorts of flowers have a good chance, the wind-fertilised Grasses, the Insect-fertilised Leguminosæ, the flowers which trust to colour, or perfume, or sweet taste—all are copiously represented in June and July. Some few indeed have hurried on their blooms to open in spring, or kept them back for autumn, as if for this small minority it were better to be out of season than to compete with the throng. Many bulbs, which can store up food when the days are long, expend part of it in flowering early or late in the year. Catkin-bearing trees flower so early because the wind can then carry the pollen to the stigmas through bare boughs instead of through leaves, which would inevitably detain and waste a great part of that small proportion which actually reaches the tree.

The botanist finds most occupation in July and August, but the great spectacle, when the woods and meadows and heaths are full of bloom, comes earlier, and is at its best on Midsummer Day.

HAY-TIME.

It is July, and in the north of England the meadows are almost ready for the scythe. The flat fields along the river look brown, as if scorched by the sun, but it is only the dull-coloured panicles of the flowering Grasses which deaden their tints. I can see from my window one patch of emerald green in the sea of brown ; it is a field which has just been cut and cleared.

Grass-pollen floats everywhere in the air. I find it on my microscopic slides, and sufferers from hay-fever find it to their sorrow in their nostrils. Dr. Blackley, when prosecuting his ingenious researches into the cause of hay-fever, found grass-pollen at considerable heights in the air. He raised two and even three kites, one above another, the lower holding the string of the one beyond it, and so was able to expose slips smeared with glycerine at elevations of several hundred feet. The spread of grass-pollen to great heights in the air, and its penetration to the recesses of our houses give proof of the extreme lightness and profusion of the grains. While almost every stigma becomes fertilised, innumerable grains are wasted. Insect-fertilised flowers waste little pollen, but they have to maintain an elaborate machinery to secure this advantage.

A friendly correspondent, Mr. B. Holgate of Leeds, tells me of the curious spectacle that may often be observed when a field of hay is cut. The wild animals which lurked in the long grass are driven towards the centre as the scythe or mowing machine works round the field from the outside. Rabbits, Field-mice, and now and then a Weasel or a Hare may be imprisoned in the ever-narrowing patch of uncut grass. In the hayfield Hares are seldom caught in this way, but when a cornfield is reaped, they are often unable to escape. The noise of the reaper

terrifies them to such a point that they will lie down
and submit to be knocked on the head. Smaller
animals are often cut to pieces by the machine. When
only a few square yards remain to be cut, the
labourers arm themselves with sticks, and watch for
anything that runs out. A good dinner is often got
in this way, but among the miscellaneous collection
of animals killed are many that no one would eat.[1]

Grasses, as every farmer knows, are of many species,
and every field contains a mixture of several kinds.
A few, among which are our chief cereals, are annual,
the majority perennial. When a perennial grass,
sprung from seed, has once established its rootstock,
sent its roots downwards into the earth and expanded
its leaves, runners are pushed out, which travel on or
beneath the surface of the ground, sometimes to a
distance of several feet, rooting at intervals and
forming fresh tufts of leaves. The runners are solid,
and often sheathed in scales, which are really a kind
of leaves. Runners which lie on the ground are
green, but the subterranean ones are blanched.

I have lately gained a practical knowledge of the
runners of one particular Grass, *Holcus mollis*, which
is, I am told, known to farmers as the Yorkshire Fog.
Our tennis-lawn was sown last year with fine grasses,
but in the old sods which formed part of the soil
were many bits of Yorkshire Fog, which soon began
to show themselves above ground. Every one must
know this grass by sight, if not by name. It has

[1] A graphic account of the disturbance of wild creatures by
the mowing of the grass is given by Cornish in *Wild England
of To-day*, p. 243 (1895).

broad, soft, pale-green leaves, and large panicles of whitish or pale purple flowers. Cattle dislike it, and it is often left quite undisturbed in a pasture, forming coarse tussocks, or even covering large spaces to the exclusion of more profitable species.

Of course the Yorkshire Fog cannot be tolerated in

FIG. 46.—Creeping root-stock of Yorkshire Fog. (*Holcus mollis.*)

a tennis-lawn, and I set to work to extirpate it. This I did with great labour because of the runners, white, horizontal stems, spreading through the earth an inch or two beneath the surface. The runners branch, root themselves at frequent intervals, and continually send up bunches of leaves. Bits of the runners are easily left behind, and these sprout again, so that whoever undertakes to eradicate Yorkshire Fog, when once it is fairly established, has his work cut out for him. There is another very common Holcus (*H. lanatus*) which is downy, and sends out no runners.

The rootstock is the true stem of the grass, and the runners are its branches; it is these which bear the buds. In summer the grass sends up flowering branches, which are often called stems too, but as these are quite different from the true stem, we want another name for them. Let us call them *haulms*; the straw of Wheat, Oats and Barley is made up of these haulms, together with their leaves and flowers.

The full-grown haulm is built up of lengths of hollow, cylindrical stalk, with knots at the junctions. At each knot is a plate of tissue which interrupts the cavity, and from the same place springs a leaf, which passes up the haulm about as far as to the knot next above, clasping it close all the way. This leaf-sheath ends in a blade, which is usually long, flat, and pointed. The sheath is very often split all along one side, the side opposite to the blade, but in some of our common grasses the sheath is entire, like the barrel of a quill. The leaves stand alternately on opposite sides of the haulm. At the base of the sheath is a leaf-knot, which is sometimes quite distinct from the knot of the haulm itself. Where the sheath and blade of the leaf meet is a transparent scale, the ligule, which ascends for a short distance in close contact with the haulm. It has been conjectured that the ligule prevents rain-water from making its way into the cleft between the sheaf and the haulm. I am not at all sure that this is the real purpose of the ligule water does not easily enter a narrow, air-filled space.

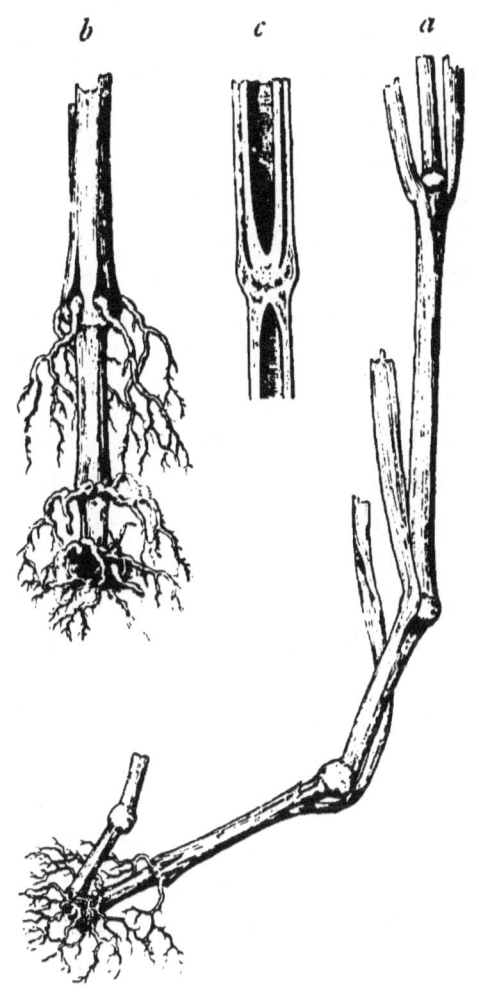

FIG. 47.—Haulm of Oat. *a*, base of haulm, showing knots and leaf-sheaths; *b*, roots of ditto piercing the leaf-sheaths; *c*, section through a knot and leaf-sheath.

When the haulm is young, it is solid, and completely filled by soft cellular tissue. The sections between the knots are at this time very short, and at first the leaves and their sheaths grow faster than the sections of the haulms, so that they greatly exceed them in length. One result is that the sheaths overlap, and a cut across a young haulm may show several sheaths, one inside another. The leaves borne on the haulm are profitable to the plant in the early season, and they attain a fair length in April or May, but the haulm itself is then quite short. It does not rise to any considerable height until the flowering-time is close at hand. Till then unusual height would bring with it the risk of laying by the wind and rain, and no corresponding advantage.

But when the flowers are ready to open it is necessary that they should clear the low herbage. The pollen, and a little later the seeds, have to be dispersed, and this is best effected when they are carried on tall, elastic stalks which dance in the wind. The haulm, which was short and succulent, now rapidly expands, shooting upwards and enlarging in diameter at the same time. A haulm of a certain grass, *Festuca elatior*, which I measured from day to day, lengthened two inches in twenty-four hours during part of the time. The growth is so rapid that the cellular tissue in the interior tears open, and the haulm becomes hollow. There is comparatively little increase of weight, chiefly expansion and hardening of tissues already formed. The sections of the haulm increase in length till they equal or a little exceed the leaf-sheaths, and these, which were telescoped one within another, now become drawn out. It is not uncommon to find the terminal joint very much prolonged, and then it stands up as a slender wiry stem, clear of all the leaves, and loaded at the summit with spikelets of flowers. The sudden expansion of the haulm is possibly the chief reason for the split leaf-sheath. The sedges, in which the haulm expands and ascends more slowly (the solid pith is one proof of this), have the leaf-sheaths closed. In some small and slender grasses, where the haulm never shoots up to any considerable height, the sheath is closed; so it is in the Cock's-foot grass, where the haulm dilates only slightly, and remains nearly solid, though here the swelling of the bulky spikelets tears open the upper leaf-sheaths.

Each section increases in length by growth at its base, and the young and tender tissues above the knots probably derive support from the leaf-sheath which wraps them round.

The knots bind the fibres closely together. They also stiffen the haulm by forming a diaphragm or floor across the tube. Some authors have doubted whether the mechanical function of the knot in stiffening the tube is of practical importance, but after examining a number of common grasses with special reference to this question, I have no hesitation in saying that the knots do materially stiffen the haulm. They also discharge a very definite and useful function of another kind. They remain capable of absorbing water from the surrounding tissues, and of swelling in consequence. In the uppermost part of the haulm, which is erect, the swelling could produce no useful effect, and here there are no knots except where the flower-stalks are given off. But at the base of the haulm the knots become crowded. Here the unequal turgidity[1] of the cellular tissue can be turned to good account. That side which comes next to the ground swells, while the opposite side becomes compressed or even folded. When a haulm has been pushed out sideways from the rootstock it is the unequal turgidity of the knots which causes it to curve upwards and take an erect position. After a storm of rain and wind has laid the long stalks, the turgidity of one side of the knots (always that side which comes next to the ground) erects it once more. The deflection capable of being produced by a single knot is only moderate, but as the knots are crowded together at the base of the haulm, a considerable aggregate effect can be brought about. Inclination through an angle of ninety degrees can be caused by the unequal turgidity of three or four knots.

Thickenings containing tissue whose turgidity can be regulated are found also at the points where the inflorescence of a grass branches. It is by means of such organs that the lateral stalks, the flowers, and even the bracts of the flowers, change their posture as required. In the bud all the flower-stalks are limp and collapsed, but when the flowers ripen the stalks take up the most favourable position, often at right angles to the haulm. Turgidity, or the want of it, regulates the exact place and attitude of every member of a highly complex inflorescence.

Additional roots are often emitted from the lower knots of a grass-haulm. They push out through the leaf-sheath, which is ruptured to give them passage.

[1] *Turgidity* is distension of the tissues by water.

In some large grasses, such as cereals, these additional or adventitious roots can be seen at times to issue from a knot two or three inches above the ground. They serve not only as channels for the supply of nourishment, but as stays, like the stout wires which are employed to secure a telegraph-pole.

I have said that the diaphragms probably increase the rigidity of the haulm ; in the great haulms of a Bamboo they certainly do stiffen and strengthen the tube to a notable extent, but here they are of particularly firm texture. An excellent model of the grass-haulm can be made by procuring one of the Bamboo rods now sold in the shops for fishing-rods and curtain-poles, and sawing it in half along its length. If a piece of Bamboo is knocked to bits against a stone, some notion may be got of its great strength, and of the way in which the diaphragms prevent splitting and crushing. The Bamboo is a true grass, and in all essentials of structure reproduces on a large scale the features of our small native grasses.

If we try to cut a piece of Bamboo with a knife we are reminded of another peculiarity of the grasses. The Bamboo cuts very badly, with a harsh gritty feel, and quickly blunts the knife. The hard, glossy surface is particularly unpleasant to cut, as well may be, seeing that it is largely composed of flint. The glossy surface of a Wheat-straw contains much flint too, and in various degrees all grasses and almost all parts of them are flinty. By very cautious charring, or by removal of the organic matter in other ways, it is possible to get small pieces of grass which show the flinty particles under the microscope. They sometimes form a uniform sheet, or in other cases rows of minute beads. The flinty covering prevents the penetration of moisture, gives additional rigidity, and perhaps defends the plant against the attacks of certain animals. Some grasses are so effectually protected that they are hardly ever eaten, but in other cases the softer tissues especially are eaten out by small, burrowing Insect-larvæ. Browsing cattle disregard such trifles as a microscopic layer of flint, but even browsing cattle can be kept off by the defences of certain common grasses. Nothing perhaps is more effectual than a close covering of fine hairs on the leaves and leaf-sheaths. These prevent easy wetting, and the leaves become unpleasant to chew. One reason why Holcus is avoided by cattle is apparently that its leaves are hairy.

How the grass-haulm is adapted to endure the wind, how its proportions secure adequate strength without waste of material, and how it comes to possess such elastic stability that it sways beneath a light breeze and yet is not prostrated by the storm, are questions whose complete solution would I believe exceed the powers of any mechanician. And the problem might be further complicated to any extent by taking into account the varying proportions of different haulms, the varying loads which they have to carry, and the minute structure of the hollow stalk, which is far from homogeneous. Nature, working by endless experiments, gives us a number of practical solutions of questions which have actually come up for settlement. She invites us to recover the question from the answer, and to compare the practical with the theoretical solution. I do not venture to accept the challenge, but shall merely look about to see whether some particular cases, artificially simplified by assumptions, can be illustrated by known results of calculation and experiment.

A joint of the grass-haulm is usually a hollow cylinder. What is the advantage derived from this form? A column of circular section is particularly appropriate to a structure which has to resist pressure from all sides. Such a column may be either a solid or a hollow cylinder. Of these two the hollow cylinder is clearly the stronger for a given sectional area. This becomes evident when we consider what will happen during bending. One side (the convex side) will be stretched; the opposite side will be compressed, and between the two there will be a neutral line, where there is neither extension nor compression. Material lying close to the neutral line would be less useful, and would stiffen the cylinder more if it were removed and disposed uniformly on the outside of the rod where the extension and compression are greatest. A hollow cylinder is inevitably stronger to resist bending than a solid cylinder of the same weight per foot run.

The joints of the haulm get narrower in regular succession upwards, the long joint which carries the spikelets being conical and usually very slender. This is attended with various advantages. The moment of the load varies with the distance, and is least at the summit of the stalk. Hence economy of material is obtained by a reduction of thickness at that point. The surface exposed to wind is reduced where the wind would act most violently. The centre of gravity is brought nearer to the ground. The haulm is stiffest where the overturning moment is greatest, most flexible where the spikelets are situated. It is important to many grasses for the dispersal of their pollen and seeds that the spikelets should dance in the wind.

The diaphragms of the haulm have, as we noticed, some effect in stiffening the structure. When a hollow cylinder is bent, the opposite surfaces tend to approach the neutral line, and the cross-section becomes elliptical. Whatever resists that change of shape, such as a solid floor, will oppose bending. The diaphragms are most crowded where excessive bending would be most injurious, *i.e.*, near the ground, It is probable that the diaphragms offer no appreciable resistance to moderate bending.

The mechanics of a long bone cannot be treated without raising some of the same questions. This subject has already been handled by Dr. Donald Macalister in a lecture which is peculiarly interesting and at the same time perfectly simple. I should merely refer the reader to his article in the *English Illustrated Magazine*,[1] if it were not that every reader has not ready access to the old volumes of a periodical.

Dr. Macalister points out that the tubular form of a long bone, such as the human thigh-bone, fits it to resist either a breaking or a crushing stress. A solid cylinder of the same mass would be weaker than the hollow cylinder. If the solid cylinder had a diameter of 100 units, and the hollow cylinder an external

[1] 1883-4, p. 640.

diameter of 125, with an internal diameter of 75, the area of cross-section would be the same in both cases. But the hollow cylinder would have a power to resist breaking greater than that of the solid cylinder in the proportion of 17 to 10, while its resistance to crushing would be more than twice as great. The strongest tube is one whose external and internal diameters bear the proportion of 11 to 5.

Dr. Macalister shows that the cancellous or lattice-work arrangement of the internal laminæ of bones coincides with the lines of pressure and tension. Hence the bone is strengthened precisely where strength is most needed, and the stresses are distributed. The bony substance is placed along the curved pressure-lines and also along the curved tension-lines, which intersect the first at right angles. But the intervening neutral spaces, where there is little or no thrust or pull, are left unoccupied, thus economising material and diminishing weight as far as possible.

In some animals whose weight is "taken off" by the water in which they live, increased weight of the body is less disadvantageous, and economy of substance may be disregarded for the sake of additional strength. The bones of a Crocodile are solid, and composed almost to the core of a dense, ivory-like substance. In a land-quadruped weight must be more carefully considered, and the long bones are largely excavated, the spaces being to a great extent occupied by marrow. In a flying Vertebrate strength and lightness are combined with still greater nicety of calculation. The wall of the shaft is reduced to a thin shell, the cancellous tissue is scanty and large-meshed, and the cavities are filled with air instead of marrow.

The limbs of many Crustacea and Insects illustrate in their way the advantages of the tubular principle. But the best example of the strength and lightness yielded by the tubular structures of animals is furnished by the hollow quill of a Bird's feather.

The leaves of grasses are full of curious contrivances, some of which are described and figured by Kerner in his *Pflanzenleben* (*Natural History of Plants*).

The flowers are expressly adapted to wind-fertilisation. Notice the absence of striking colour, scent or honey, the abundance of the pollen, the lightly poised anthers, and the feathery stigmas.

When the fruits ripen and fall off, there fall off with them certain of the enclosing husks. It is only in some cultivated cereals that the artificially enlarged grain can be readily detached from its envelopes. It is not uncommon to find the flower-stalks jointed, so that they readily break away from the haulm. The husks serve to protect the grain from spoiling by rain or drought, and in some cases aid in dispersal by greatly increasing the surface exposed to wind. My old enemy, the Yorkshire Fog, enjoys great facilities for dispersal by wind, and I find it springing up in the most unlikely corners of the garden. In the Reeds of the fenlands some of the inner husks are fringed with long, silky hairs, which act like the hairs on the seeds of Willow, wafting the grain to long distances. Some grasses have awns attached to the husk, which catch in the fleece or fur of animals. In

Feather-grass the extremely long and feathered awn twists when dry, and untwists again when wetted, thus screwing the pointed fruit into the earth, the long awn, entangled in the herbage, furnishing a fixed point to push against.

Small as it usually is, the grass-fruit carries with it a little store of starchy food and a minute quantity of a ferment, which, under suitable conditions of moisture and temperature, dissolves the starch, and renders it fit for assimilation by the embryo plant.

A meadow ripe for the scythe calls up before me the endless contrivances by which the grasses have won such mastery in the struggle for the surface of the earth. But what different thoughts the same sight may suggest to other minds! Andrew Marvell, walking behind the mowers at Nun Appleton, was chiefly struck by their resemblance to the Israelites passing through the Red Sea!

> "Who seem like Israelites to be
> Walking on foot through a *green* sea,
> To them the grassy deeps divide,
> And crowd a lane to either side."

The History of the Cabbage White Butterflies.

When I came to live in the country I naturally began to grow cabbages. One result has been that I have great facilities for the study of Cabbage Whites. In May I find the eggs on the leaves; the eggs hatch out and produce caterpillars, which are too plentiful for any but naturalists during a great part of the summer. The caterpillars turn to yellow pupæ, spotted with black, which are found on the

FIG. 48.—Large Cabbage White Butterfly; the female above, the male below.

trees, walls and palings around the cabbage-plot; and from these pupæ issue white butterflies, such as I used to chase across the summer fields in my schoolboy days.

When we look closely at the White Butterflies,

which are reared in our kitchen garden, we find that they belong to three different species, of different sizes, and with rather different markings. These are distinguished as the Large, the Small and the Green-veined Cabbage Butterfly. If you find any common Cabbage Butterfly and wish to identify it, you may find it convenient to use this table of characters.

	Size across expanded Wings.	Fore Wings.	Hind Wings.	Larva.
Large Cabbage White.	2½ in.	*Male.* Black at the tip; no spot. *Female.* Black at the tip; two black spots and a dash.	A black spot.	Yellow, spotted with black.
Small Cabbage White.	2 in.	*Male.* Blackish at the tip; no spot or a blackish spot. *Female.* Blackish at the tip, two black spots and a faint dash.	A black spot.	Green, spotted with black and yellow.
Green-veined Cabbage White.	1½ to nearly 2 inches.	Blackish at the tip; one black spot.	*Male.* A black spot or none. *Female.* A black spot. Under side with greenish veins in both sexes.	Green, spiracles red, on yellow spots.

The larvæ feed upon Cabbages and allied plants (Cruciferæ). The small and green-veined species feed also on Mignonette and Tropæolum. There are two and sometimes three broods in the year. Eggs laid in the end of summer yield larvæ which pupate in autumn, and after hibernating emerge as Butterflies in spring. The Butterflies lay eggs, and the issuing summer-larvæ pupate about midsummer. In an early season the broods are hastened a little, and the

FIG. 29.—Small Cabbage White: the female above, the male below

autumn brood is the third and not the second of the year. From April to September there are few weeks in which the Butterflies cannot be seen, and successive generations of caterpillars are busy feeding during most of the summer.

The eggs of the Large Cabbage White are laid in patches on the underside of the leaves, and are small yellow objects. Keen eyes will now and then detect he eggs (hard boiled) on cooked cabbage leaves. If you capture a female Butterfly, she will probably lay eggs for you, but the best way of securing a supply of eggs is to watch a female Butterfly as she haunts the

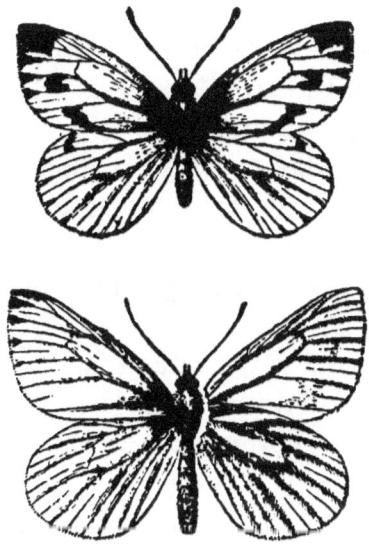

FIG. 50.—Green-veined Cabbage White ; the male above, the female below. In the lower figure the under side of the wings are shown on the right.

Cabbages. When she rests for awhile among the leaves, mark the place. After she has flown away, you will often find that she has left a hundred eggs or so on a leaf. The small Cabbage White lays her eggs singly.

Why are the eggs laid on the *under* side of the leaf? Perhaps to protect them from sun and rain ; perhaps to keep them, or the larvæ which issue from them, out of sight of greedy animals. The female Butterfly is not very scrupulous about the position of the eggs, and will sometimes lay them on the upper surface of the leaf.

The eggs, when examined by a lens, are seen to be enclosed within a flask-shaped cell, on which is a delicate relief pattern. The egg-shell is quite impervious to liquids. I once wished to mount a number for microscopic examination. Knowing that the larvæ, if not killed, would by and by emerge and break the shell, I resolved to soak the eggs in strong alcohol for some days. This was done, and then the eggs were cemented upon a glass slip. But they hatched out all the same, and all the egg-shells were destroyed. Plunging the eggs into boiling water is a better expedient.

In ten days or less the caterpillars hatch out. They have the outward form usual in Lepidoptera. There is a dark-coloured horny head bearing the jaws (chief

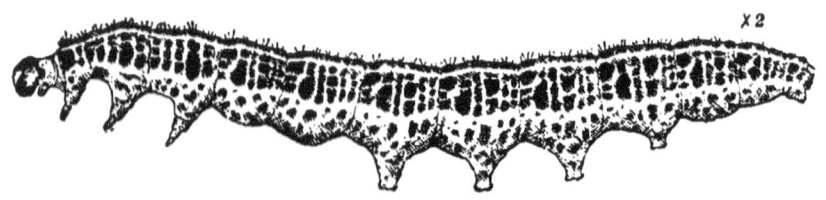

FIG. 51.—Larva of Large Cabbage White, × 2.

among them the powerful mandibles), a pair of eye-spots, and a minute pair of antennæ. All these require magnifying power for convenient observation. There are three pairs of thoracic legs, and further back four pairs of prolegs, besides a pair of claspers at the hinder end of the body. Except where reduced for special reasons, as in Geometer larvæ, the legs of all Lepidopterous caterpillars have the same number and disposition.

When I have examined the outward appearance of an Insect, I like to anatomise it, and here comes the chief interest. Just as, in the words of Tony Lumpkin, the inside of the letter is always the cream of the correspondence, so the inside of the Insect is the best part of its structure. I should like very well to talk about the things which can be seen in a caterpillar by the dissecting microscope, the air-tubes, the nerve-cord, the heart, the digestive tube and the

FIG. 52.—Larva of Small Cabbage White, × 2. After Buckler.

reproductive organs. It is worth while to note that fresh-hatched larvæ are already male or female. During the whole larval period the reproductive organs slowly increase in size and complexity, and when the change to the pupa takes place, they are often, to the eye of the anatomist, perfectly formed. But this is hardly the place to describe in detail things which can only be followed with the scalpel and lens, and I will say no more about the anatomy of the larva.

The larva changes its skin four, or sometimes five times, the last change being that which we call pupation. Then the last larval skin is cast, and the pupa is disclosed. In the preceding changes of skin little alteration of form, only increase in size, is to be observed. The last moult, however, appears suddenly to convert the larva into a new being.

I say "appears," because, as we now know, the pupa is merely the larva in a new form. Pupation is a change of skin, accompanied by an unusual amount of change of form. When the last larval skin is cast, the rudiments of wings and other new parts become visible, though they do not acquire their ultimate structure nor serve any useful purpose until the resting-stage is over, when, after one more moult the winged Insect emerges.

This seems plain enough to any one who observes for himself. Yet much controversy was needed and much Christian ink had to be shed before men could be persuaded to drop their theories and look the facts in the face.

When our Royal Society was founded the wildest notions were abroad as to generation, development and transformation. If the reader should chance to come across a curious but rather worthless book published in 1634, Moufet's (or Mouffet's) *Theatrum Insectorum*, he will find in the dedicatory epistle by Sir Theodore de Mayerne much learned trash about the universal spirit which fills and governs the three kingdoms of Nature. If animals and plants undergo transmutation, Mayerne does not see why it should be impossible for metals to do the same. In 1651 Harvey, our great Harvey, published his treatise on Generation, and here, as all the world knows, is solid matter, the fruit of observation and reflection upon the development of chicks and fawns. But

Harvey had studied the scholastic philosophy too, and he treats us to many pages of learned disquisition. Thus he explains to us that there are two ways in which things come into being. The material may be ready to hand, and require only to be supplied with form. The sculptor makes a statue, but he does not make the marble. In other cases material as well as form has to be supplied, or at least brought together, as when a potter gathers clay, and adds bit to bit to make an image. In some animals, Harvey goes on, the material is all collected beforehand, and only requires to be thrown into shape; that is *metamorphosis*. In others the parts have to accrete substance to themselves and grow; that is *epigenesis*. Insects are developed by metamorphosis, but the higher animals, which have blood, develop by epigenesis.

Such was the kind of speculation which was current in the learned world when Malpighi and Swammerdam began to explore the transformations of Insects with the scalpel and the microscope. I do not know which was the first to observe the fact, but Malpighi was the first to announce that in a Lepidopterous larva nearly ready for pupation the legs and wings of the imago may already be distinguished by dissection. The observation is to be found in that memorable treatise on the Silkworm, which Malpighi wrote for our Royal Society in 1668, and which they printed in the following year. When the larva, he tells us, has spun up, its skin splits, and the pupa emerges like a new animal born of the old one. The antennæ stand out in the place formerly occupied by the mandibular muscles. The legs of the Moth appear inside the thoracic legs of the larva, and the wings project from the sides of the segments, in places which were marked shortly before pupation by purple tracts. The appendages, when they first emerge, are slimy, and cohere during drying, so that before long the body and appendages of the pupa seem to be invested by a common envelope. Even before the larva begins to spin, says Malpighi, the rudiments of the wings can be made out beneath the skin of the second and third segments, while the antennæ are already formed within the larval head. The pupa is a mask, which protects and conceals the future Moth until it has grown firm and fit for the emergencies of a free existence. Malpighi rarely makes a controversial remark, and here he offers no comment on the views of the schoolmen, but quietly states the facts as he knew them to be.

In the very same year (1669) appeared Swammerdam's *General History of Insects*, a precursor of the *Biblia Naturæ*, far less complete and valuable than that great monument of industry and sagacity, but a noteworthy book which had its results. The History appeared a little later than Malpighi's *Silkworm*, as we see from the fact that Swammerdam quotes with high praise that very passage of Malpighi's which I have condensed above. Swammerdam in his *History of Insects* figures Daphnia, the Louse, the Dragon-fly, the Gnat, Stratiomys, Anthomyia, the Ant, the Vapourer, and the Cabbage-White, giving the transformations of such as undergo transformation. The text is meagre as compared with the later descriptions of the *Biblia Naturæ*, and is largely occupied with a discussion on what we now call the Transformations of Insects. Swammerdam protests with all his might against such words as Transformation and Metamorphosis. They sounded in his ears as Transmutation might sound to us, calling up such possibilities as the change of men to wolves, the change of the flesh of oxen to bees, the change of putrefying plants to caterpillars, and the change of lead to gold. Very likely Mayerne believed in all of them ; Harvey certainly believed that Insects could be generated spontaneously from putrefying matter. Metamorphosis was with him not a particular kind of growth, but an alternative with growth. No wonder that Swammerdam should bitterly remark, after giving a long extract from Harvey, that it contained nearly as many mistakes as words ; no wonder that he should hate the words about which such rank misconceptions had gathered. He insists time after time that an Insect grows in just the same sense as a plant or a Frog. We, who have not had Mayerne or Harvey to refute, wonder a little at his vehemence, and see no reason why we should not employ the very convenient terms Transformation and Metamorphosis. They mean to us, who are happily unencumbered by the rags of scholasticism, nothing more than conspicuous change in the form and mode of life of an animal. The change may be apparently sudden, as when a larva becomes a pupa, or a pupa an imago ; again, it may be insensibly slow, as when a Tadpole loses its tail and gills, and acquires legs, taking three or four months to accomplish the transition. Whether sudden or gradual to the eye, the change is always in reality gradual; the new organs are slowly developed and grow by imperceptible degrees. When the animal becomes to outward appearance transformed in a night, it is because the new parts have been concealed by the old skin, and are suddenly revealed.

Malpighi and Swammerdam knew all this as well as we do. Swammerdam tried to make it plain to his contemporaries, but in spite of his earnestness he only succeeded in a moderate degree, so much were men's words and thoughts entangled with the mischievous theories of bygone ages.

The reader will be glad, I dare say, to quit these ancient controversies and get back to matters of fact. We have seen that Malpighi and Swammerdam nearly at the same time discovered the rudiments of the imago within the caterpillar. The following instructions are based upon Swammerdam's method of procedure. Take a full-grown larva of any Moth or Butterfly which has ceased to feed, kill it with ether, tie it down with thread, and dip it several times in boiling water. The outer skin will then peel off readily, and the Butterfly will be exposed to view. I have repeatedly done what Swammerdam recommends, and have seen what he describes. When the larval skin is removed, we find a flabby, pale-coloured object left behind, which has two pairs of short and crumpled wings, three pairs of legs folded beneath the thorax, a pair of long antennæ, and a pair of long and slender mouth-appendages. Except that they are soft, wrinkled, and somewhat undersized, all these parts agree perfectly in outward form with the wings, legs, antennæ and maxillæ of the Butterfly. The maxillæ are intended to cohere by innumerable hooks, and so to form the long suctorial proboscis, by which the Butterfly will search the depths of nectar-bearing flowers.

Swammerdam came very near the truth in his positive statements about the tranformations of Insects, but his knowledge, like that of every first explorer of a very difficult subject, was incomplete on many points. Having found that the caterpillar just before pupation encloses what may be called a Moth or Butterfly, he concluded that the Moth or Butterfly had been there from the first, and that no more important change was involved than the expansion and at length the liberation of the imago. We now know that much goes on of which Swammerdam had no notion. The organs of the imago are not all present from the first. The rudiments of the wings form very early, even before the egg is hatched, but the antennæ, the mouth-parts, and the legs of the imago are formed after the last larval moult. Moreover, there is destruction of old parts as well as formation of new ones. The muscles of the larva, the silk-glands, and various other parts which are not required after the larval stage has come to an end, disappear altogether. The organs which are external, and belong to the outer cuticle, are simply cast at pupation, but what becomes of the internal organs which are no longer wanted? How do these disappear?

They are eaten up and converted into granules, which serve for the nutrition of the rapidly growing organs. Certain wandering cells, very like the colourless corpuscles of human blood, do the work. The corpuscles may be found sunk in the tissues which they are devouring, and bits of striated muscle, plainly recognisable under the microscope, have been seen buried in the protoplasm of such corpuscles. In the same way the muscles of a Tadpole's tail are eaten up by wandering corpuscles, which carry the substance which they have appropriated into the blood. How they part with it, and how the growing organs get the benefit of the food contained in the corpuscles, are questions to which we can at present give no satisfactory answer.

What is a pupa? I have found few, even among professed naturalists, who could give a full and accurate answer. The common notion is, I believe, that the pupa is a resting-stage, during which the imago or winged Insect is formed. The form of the pupa is supposed to be merely protective. Within the hard, usually dark-coloured, and therefore inconspicuous pupa-skin, the imago is believed to form.

There is some truth in this, but it is not the whole truth. Wings, legs, antennæ, proboscis, and other characteristic members of the imago, form, as we have seen, during the last larval stage. They become free for a short space at the time of pupation, but are then folded against the breast and glued down. The pupa is to external appearance a Moth or Butterfly which has glued down its half-expanded appendages ; it is enveloped in a close-fitting skin, which will be cast when the imago emerges.

Swammerdam must have often asked :—Since the parts of the Butterfly are plainly to be seen within the larval skin, how is it that the Butterfly, complete in all respects, does not at once issue from the larval skin? Why should a pupa-stage be interposed? I have not found Swammerdam's answers to these questions. It is not likely that he was able to answer them fully, for minute investigation of the tissues is requisite, and histology was wholly undeveloped in Swammerdam's time. Nowadays the most obvious course is to cut transparent sections through the organs of the Butterfly, after the larval skin is stripped off, and by microscopic examination we soon arrive at one significant fact. The organs of the Butterfly, though recognisable and externally pretty complete, are merely the outward shapes of what they will afterwards become. The muscles, nerves, air-tubes, and other histological elements are either absent or extremely imperfect. Much internal growth has to be accomplished before the wings are fit for flying, or the legs for running. This is the proper office of the pupal-stage of Lepidoptera, to carry on and complete the formation of new parts, necessary to the flying Insect, which were merely blocked out in the larva.

It is quite conceivable that the whole growth, both external and internal, might have been completed while the Insect was still to outward appearance a mere larva. Indeed this is very nearly what happens in certain Dipterous Insects, such as Chironomus. In them all the details of the future imaginal organs are, with some slight exceptions, completed in the larval stage, and the pupal stage, which lasts a very short time, often only two or three days, is employed in giving the parts the firmness which they will require, and in filling the new breathing-organs with air.

There are great differences between Insects as to the amount of structural change which goes on during the pupal stage. In some (Blow-fly, etc.) the whole larval body is at this time reconstructed; in others (Moths, Butterflies) the new parts, fashioned during the larval stage, are completed internally during the pupal stage; while in a third case (Chironomus, etc.) most of the new parts are already complete, inside and out, when pupation sets in, and require little more than to be hardened, or in the case of certain organs to be exercised a little, before they enter into full activity.

It is likely that primitive Insects never acquired wings, and led much the same life during the adult, reproductive stage as before. There are still some few Insects (Spring-tails, Silver-fishes) of which this is true. But it is common for adult Insects to fly. Flight gives facilities for finding a mate not too closely related, and for laying eggs in likely places, which would be inaccessible to an animal which could only crawl or run. Flight does not of necessity bring about any change of food. If there is no change of food, there is no absolute need of a resting-stage. The Dragon-fly feeds upon live Insects, as it did in its earlier aquatic condition; it has no resting stage at all. But the flying adult is likely to profit by a change of food. The larva, as a rule, is voracious; it needs a capacious stomach and stout jaws, but no extraordinary nimbleness or quickness of perception. When the Insect comes to take long excursions in the air, it will in general require a lighter and more nutritious food, such as the nectar of flowers. Change of food naturally brings about changes in the mouth-parts. Short and powerful mandibles will be discarded, and replaced by a suctorial proboscis. Change of mouth-parts inevitably means cessation from feeding, and almost inevitably cessation from work and travel. Then we get a resting-stage. When once established, the resting-stage may be turned to good account in refitting the internal organs, and indirectly in promoting specialisation of the earlier and later stages. Where a pupa-stage is provided, the larva may be yet heavier and slower, the fly yet more swift and light. If the flying adult is specialised, and the female capable of flying far and scenting food at a distance, the better will be the provision made for the young larva and the less the exertion demanded of it. But the more inert the larva, and the greater the interval between it and the active, quick-witted Fly or Moth, the more complete will be the change to be undergone in the resting-stage.

Though some Insects and not others are described as undergoing transformation, the essential and primitive feature is the periodical change of skin which occurs in all Insects. In many cases advantage is taken of the change of skin to secure a change of form. The interval between the last moult and the last but one, when passed in outward inactivity for the purpose of effecting conspicuous change of form, is what we call the *pupa-stage*. The more the larva resembles the imago, the less the need of a true pupa-stage. Difference of food in the early and final stages, scattered food in the larval stage, are among the reasons for conspicuous difference between the larva and the winged Insect, and indirectly reasons for a resting stage. Similarity of food in all stages of growth, and abundance of food, which is easily found and easily appropriated, are circumstances which render a resting stage less necessary.

We must now turn back and study the formation of the new organs beneath the larval skin. Many larvæ have to be sacrificed, and innumerable sections examined to make out the whole story, but we shall be satisfied here with learning the general plan of development. At the time of the fourth moult there is no indication of parts differing from those of the larva. But as soon as the last larval skin but one has been cast, a new skin, which we shall name the *pupal skin*, begins to form beneath the last larval skin. The new skin is not exactly moulded upon its predecessor, but pushed inwards here and outwards there. Where considerable prominences are to form, the infoldings are deep, and from their innermost extremities outward-directed folds project, which are shaped in some cases like glove-fingers, in others like pockets. These hollow folds are wings, legs, antennæ and other appendages, telescoped into the interior of the body, because the corresponding parts of the larva are not large enough to contain them. They are often much bent and crumpled, but in a methodical way, as the perfect symmetry of the two sides of the body shows. Either from the first or after a short interval, a second skin forms within the pupal skin; this is the *imaginal skin*. As the imaginal skin and its complex folds develop, the pupal skin ceases to grow. It is not cast or ruptured at present, but it becomes a mere passive envelope, which takes accurately the form impressed upon it by the growing parts within. The larva in the latter part of its history has accordingly three skins, one outside another, larval, pupal, and imaginal.

Insects furnish examples of every degree of complication of such imaginal folds. Where no change of form is to be effected, the imaginal skin is closely moulded upon the larval skin. Slight changes in length of leg or mouth-parts are readily brought about either by wrinkling of the new integument, which becomes extended as soon as it is freed, or by shallow infolding. Many intermediate cases of various complexity occur in different organs or in different Insects. The maximum of complexity is found in the Blow-fly and other Insects of the same family. Here the complete want of correspondence between the structure and mode of life of the larva and the fly, together with the high and special development of the organs of the fly, have led to an extraordinary elaboration of the imaginal folds, which are numerous, intricate and deep. Closely connected with this complexity of the new growth is the completeness of the resting-stage. The pupa has no external mark of a living thing; internally it is at one time reduced to simple elements, and consists of a kind of pulp, except for the undeveloped imaginal folds. Protected by the hardened larval skin, which forms a firm smooth capsule about it, it goes through the evolutions which are to transform a sluggish and voracious larva, destitute of limbs and almost destitute of senses, into a swift and adroit fly.

We now resume the history of the Cabbage White. I will next quote Réaumur's account of the method of fixation of the pupa. His pleasant and leisurely narrative easily admits of condensation, and I propose to condense it greatly. Those who prefer the freshness of a discoverer's narrative should read for themselves the ninth, tenth and eleventh memoirs of the first volume of the *History of Insects*.

Réaumur first tells us of Butterfly larvæ, which like those of the Peacock and small Tortoise-shell, suspend themselves at the approach of pupation by the tail, and hang head downwards. The larva spins a web upon a leaf or other support, crowding the threads towards the centre, so as to form a projection or hillock. This web is not easily seen, but it can be made evident by placing a larva ready to pupate in a box lined with black paper. To the hillock the larva applies the hooked claspers at the end of its body, and so gets a safe attachment. The next thing is to cast the larval skin. The soft parts within, covered of course by the thin and flexible pupa-skin, are swollen and contracted by turns until they become loosed from their envelope. The tail-segments are the first to be disengaged; afterwards the fore part of the body is powerfully distended, and the larval skin cracks longitudinally behind the head, allowing the surface of the pupa to appear. The next step is to slip the loose skin backwards over the body. This is accomplished by the successive contraction and dilatation of the segments one by one, and is aided by spines or backward-pointing hairs upon the surface of the pupa, which act like a ratchet, and prevent the disengaged integument from slipping back to its former place. When the old skin, like a stocking pushed down the leg, is gathered into a mass of folds close to the extremity of the body, the Insect proceeds to free itself completely, and further detaches the cast skin, which if allowed to remain hanging by its side, would needlessly attract the attention of Birds. How is a pupa, hanging by its tail, and without means of holding on by the rest of its body, to attach itself anew, and dislodge the cast skin? The tip of the abdomen of the pupa bears a pair of prominences which are opposable and armed with many small hooks. At their base the abdomen is indented, and forms a kind of elbow, which can be flexed, and used as a means of grasping. The pupa extricates the tip of its abdomen, using the elbow and the hooked forceps alternately as a means of attachment; it then creeps a short distance along the cast skin, and gets an independent hold of the hillock of threads. Next it sets its body spinning, first in one direction, and (if necessary) in the opposite direction by turns. The hooks cut through the threads which hold up the cast skin, and this falls to the ground.

Where the pupa is to hang by its tail, head downwards, the artifice of the Peacock and Small Tortoiseshell Butterflies answers perfectly, but the Cabbage Whites and some others have reasons of their own for taking a more or less horizontal position, or if placing themselves vertically, keeping the head uppermost. Here a second attachment becomes desirable, and they secure themselves by a girdle passing round the body well behind the head. Réaumur goes on to describe the girdled pupæ.

Each end of the girdle is glued to the supporting object. At first sight the girdle looks like a single thread, but on examination with a lens it is found to be made up of many threads, which are neither glued together nor interwoven. It is sufficiently loose to allow the body to move a little way in any direction, and, what is of special importance, is loose enough to allow the cast skin to be slipped off beneath it.

The larva attaches itself by its claspers to a hillock of silken threads, and remains quite still for many hours before beginning to spin the girdle. The subsequent operations differ a little in different Butterflies. Réaumur kept several kinds in captivity, and was rewarded by discovering three modes of procedure, each adapted to the wants of a particular species.

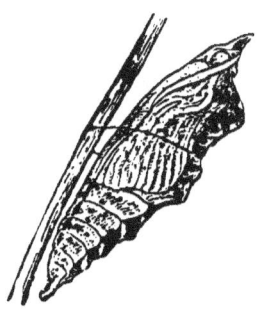

FIG. 53.—Pupa of Small Cabbage White. × 2.

In one of the Hair-streaks (Thecla) the larva is stumpy and covered with stout hairs. Having previously attached itself by its tail, it contracts the forepart of its body as much as possible, fastens the beginning of a new thread to the support, and passes it over its head to the other side of its body, where it fastens it again. The head is employed in a peculiar way to carry the thread across. Some notion of the process can be got by holding a thread between the finger and thumb of the left hand, and grasping it again with the finger and thumb of the right hand. Then the right finger must be turned so that the thread rests upon its nail, which answers to the hard and shiny surface of the head of the larva. By this extempore model the reader can better understand how the thread issuing from the mouth is made to sweep across the body in an arc of the requisite size. gliding smoothly all the time over the polished head Each thread as it is fixed is passed backwards over the spiny segments, which contract or dilate for the purpose of aiding its passage, and thus by the addition of many threads the girdle at length acquires due strength. Then the fore part of the body is extended, the head passed well in front of the girdle, and the larva, now provided with a two-fold attachment, can proceed to divest itself of its larval skin without fear of falling to the ground.

Réaumur tells us next how the Large Cabbage White manages. When pupation approaches the larva makes its web and hillock, as already described, and catches hold with its hooked claspers. The body is smooth, and so flexible that the head can be bent backwards and made to touch the fifth segment. Thus doubled up it spins the girdle from side to side, passing it round the furrow between the fifth and sixth segments of the body. When sufficient silk has been spun, the body is straightened, and comes into the best position for support by the girdle.

The third method of girdle-spinning is practised by the Swallow-tail Butterfly. Here the larva holds on by its tail and abdominal feet, the head is thrown well back, and the thoracic feet are in the air. The thread, as it proceeds from the spinneret, is caught by the fore legs, and held taut, as by the fingers of a person holding a skein of wool. When the girdle is completed the larva slips its head through, and is at once adequately supported.

In all three methods it is requisite that the girdle should surround the body at a point well behind the head. The tail is already fixed, so that the body cannot be moved forward as a whole after the girdle is completed, but by one or other of the three methods described, viz.: (1) contracting the fore part of the body during spinning; (2) doubling it up; (3) arching it away from the fixed support, the girdle is set far enough back, and a due amplitude is insured.

When the pupa first becomes exposed, it is much like the pupa liberated artificially from the larval skin, and has in essentials the same external form as the future Butterfly. The limbs and antennæ and proboscis are separate for a moment. Then they are gently drawn over the breast,[1] the proboscis in the middle line, and the others in perfectly regular pairs outside it. A viscid fluid is poured out, which sets on exposure to the air, and glues them fast. After this the pupa can only move its abdominal segments, and even this it rarely does, except when disturbed.

During the pupal stage there are no outward signs of life, although considerable internal changes are in progress. The alimentary canal becomes smaller and more complex; a sucking stomach is partitioned off from the larval crop, for use in drawing up nectar

[1] In some cases this appears to be effected directly by the process of extrication.

through the proboscis; the nerve-cord and dorsal vessel become shortened; the silk-glands practically disappear; the reproductive organs enlarge; the new appendages acquire their motive and sensory apparatus; the voluminous fat-body of the larva is used up. No food can at present be taken into the body, but the pupa breathes all the time, and perishes if its supply of air is cut off.

At last the organs of the Butterfly have attained the last degree of perfection; the pupal skin cracks along the back of the thorax, and the winged imago emerges. At first its wings are damp and crumpled, but they speedily expand and stiffen, and in no long time the Butterfly is ready to range the fields, seek its mate, and provide for new generations.

Not all the larvæ bring their life-history to a prosperous end. There are some, in particular years a large proportion, which are attacked by a deadly enemy, an Ichneumon fly (*Microgaster glomeratus*), which pierces the skin, and lays its eggs in the living body. The eggs hatch, and the larvæ which issue from them devour their host alive. The victim has not strength enough to assume the pupal stage. It creeps up some adjacent object, as if with the intention of casting its larval skin, but remains immovable, and unchanged. The parasites now devour all the viscera, creep out from the empty skin, and keeping together, spin each its own little cocoon of yellow silk. A cluster of such cocoons may often be seen hard by the empty larval skin, and people have been known to take them for the eggs of the Cabbage White and destroy them. The fully developed Ichneumon is a small four-winged fly, with piercing ovipositor, or egg-shoot, and long antennæ, which vibrate rapidly, as if to gain from the surrounding air some intimation of the neighbourhood of their prey.

There is, in this as in other like cases, a peculiar relation between the abundance of the caterpillars and the abundance of the parasites, which is only fully brought out by long-continued observation. Let us take as the first of a series of years one in which the caterpillars are plentiful and the Ichneumons few. This state of things favours the increase of the Ichneumons. The caterpillars become infested in large proportion, few pupæ yield Butterflies, many yield Ichneumons. When things come to a climax the Ichneumons are extraordinarily plentiful, but their victims fewer than usual. Then the Ichneumons suffer from mutual competition, and many die without propagating their kind. As their numbers decline, the numbers of the caterpillars increase. So the cycle comes round time after time, the maximum of the Ichneumons lagging behind the maximum (and in some cases nearly coinciding with the minimum) of the caterpillars.

CABBAGES AND TURNIPS.

My cabbage-plot is, I must admit, ridiculously small, but it gives me plenty of opportunity for observation. The difficulties of the young plants in dry spring weather, and their rapid growth in a wet June are among the little events of our year. The way in which the leaves throw off rain, gathering up the moisture into big drops, which roll to the earth about the roots instead of choking the breathing-pores, and the waxy bloom which brings this result about, are well worth both study and admiration. I hold with Andrew Fairservice that a kail-blade by moonlight is like a lady in her diamonds. Many of the drops which we call dew-drops are not deposited upon the leaves by condensation of vapour, but exuded as liquid. You can see this very plainly in the Cabbage, for the drops appear at certain points only on the margin of the leaf, where veins end. Here are special water-pores. Drops are exuded whenever the tissues of the plant are chilled, most abundantly when a cold night, whether clear or cloudy, follows a hot, damp day. True dew-drops form only beneath a clear sky, and condense as a multitude of minute globules, which may afterwards roll together. The mere cabbage-stalk is a wonder in its way if carefully examined. Look out for cabbage-stalks which have been thrown aside to bleach in the rain and sun, not in the filthy air of a town, but on a country farm. You will see the stout network of fibres, the meshes which allow the cellular tissues to expand and to communicate with one another, the bundles of vessels which pass to the roots and the leaves. If you can get a turnip bleached in the same way, compare the two, and notice that the turnip is merely a bulge upon what is essentially a cabbage-stalk too, though it is here called a root. There is plenty of occupation, to say nothing of provocation, to be got out of the Insects which haunt a cabbage-ground. But the chief interest of cabbages, turnips and the like is to me the part which they have played in human civilisation.

FIG. 54.—Wild Cabbage. From Sowerby's *English Botany*.

Here and there along the south coast of England and the Welsh coast we find a plant known as the

Sea-cabbage (*Brassica oleracea*). It grows in tolerable plenty on the chalk cliffs of Dover, and is also recorded from the Isle of Wight, Cornwall, South Wales, and Great Orme's Head. It is about twenty inches high. The leaves are large, jagged, and covered with a blue-green bloom. The stem is tough and woody. The flowers are of a pale yellow colour, and are succeeded by pods. The plant belongs to the order of Cruciferæ, the same large and important order which yields the Radish, Mustard and Watercress.

From this wild original (or possibly from it and one or more closely allied forms not easily distinguished) have been derived the countless varieties of the cultivated cabbage. Red cabbages, Brussels sprouts, with their crowds of little leaf-buds, cauliflowers, with their dense masses of imperfect flowers, brocolis and savoys, are all cultivated forms of the weedy and ragged sea-cabbage. In Jersey, Mr. Darwin tells us, a cabbage-stalk has grown to the height of sixteen feet, and has had its top occupied by a Magpie's nest, while the woody stems are often ten or twelve feet long, and have been used as rafters and walking sticks. A cabbage-stalk fashioned into a walking-stick may be seen in the Museum of Economic Botany at Kew. The principal varieties were established before botanical curiosity had been excited, and we can only get chance bits of information as to the time and place of their first appearance. Theophrastus knew of three cabbages, Pliny of six. Regnier has collected evidence that cabbages were cultivated by the Celts of ancient Gaul. There is no hint that they were known to the ancient nations of the East, and De Candolle, who made laborious researches into the subject, believes that the cultivated cabbage is of European origin.

Turnips are practically cabbages in which the lower part of the stem, beneath the seed-leaves, has become enlarged and fleshy under cultivation. Botanists think that the wild turnip and cabbage, though extremely similar in form and mode of life, are capable of separation, but this is a question for specialists and of little practical moment.

Cabbages and turnips yield striking examples of conspicuous changes due to long-continued cultivation and selection. They must have been factors of appreciable weight in the early civilisation of Western Europe. We can imagine some old European savage, wandering dinnerless along the seashore, until at length he was pressed by hunger to experiment upon unfamiliar plants. That savages do thus gain knowledge at the risk of their own lives we may infer from the well-known fact that they are well acquainted with the properties of the common plants of their own country, and can point out which are poisonous, which useless, which good for food. Our savage sees the tall, weedy sea-cabbage, and finding nothing more tempting, tries its flavour. There is a slight pungency of taste, which raises misgivings, but no ill-effects follow. Next day the sea-cabbage is again resorted to, and in time becomes a regular article of food. Presently some ingenious fellow, the Watt of his age, saves himself the trouble of a daily journey to the shore by transplanting a few cabbages to a patch of ground near his cave. The refuse which lies around, unwholesome as it is to men and animals, encourages the cabbages to more vigorous growth. Years, perhaps centuries later, another great advance is accomplished, and men begin to raise the cabbage from seed. Gardens and fences follow. It is no longer necessary to spend whole days seeking food, and the man's hands are set free to make himself shoes, and a coat, and a house.

We have perhaps given to the cabbage some share of the credit which rightfully belongs to barley or some other nutritious plant, but there is no doubt that the cabbage played a considerable part in the early civilisation of Western Europe. Cultivated plants and domestic animals are the very foundation of primitive society. As the plants grow more juicy, and the animals more docile, Man too rises to something higher than he was. He becomes able to lead the life which pleases him, and not that which is imposed by climate and the wild productions of the soil. He learns by slow degrees to shape his own circumstances and habits. But his intellectual gifts and his social aptitudes cannot be developed without certain simple natural resources. Of these the chief are plants worth cultivation and animals worth domestication.

We have no distinct record of the time when cabbages and turnips were not cultivated in Western Europe. But until modern times they were cultivated in gardens, by the spade, and on a small scale. No doubt the first cultivation of vegetables in gardens, could we get to know all about it, was the important

CABBAGES AND TURNIPS 189

step, but the mere multiplication of useful vegetables by wholesale culture had great effects upon the health and prosperity of the people, and this part of the story admits of being set down in some detail.

As late as the time of the Civil War the cattle and sheep of England had to endure something like starvation every winter. Between harvest and ploughing the unenclosed arable lands were used in common for grazing, and formed together with the pasturage of wastes and moors, the chief subsistence of the flocks and herds. Hay was made in small quantity, for the ground, was unfenced, and no diligence of the haywards could keep the animals from devouring or treading down the long grass. At the approach of winter all the livestock was killed and salted, except such as were kept for breeding. No grasses were raised from selected seeds till the eighteenth century, though clover and other " artificial grasses" had been introduced from the Low Countries a century earlier. Turnips are said to have been brought over by Sir Richard Weston. He had been ambassador at Brussels (1620-2), and when he came back he cultivated turnips and artificial grasses in fields at Sutton in Surrey.[1] At this time the Flemings and the Dutch were the most advanced of European nations in horticulture and agriculture, and their vegetables and seeds were largely imported by England. The English engineer, ship-builder and merchant of that age looked to the Dutch for

[1] *A Discourse of Husbandrie used in Brabant and Flanders.* London, 1650. 4to.

example and training just as naturally as did the few English farmers who dreamt of adopting improved methods. The turnip, which Tusser[1] had called a "kitchen-garden root to boil or butter," was slowly taken up as winter-food for sheep. Blith (1652) derides turnips altogether, and says that even swine will only eat them when boiled. Jethro Tull claimed to have raised turnips in the field in King William's reign, but he adds that "the practice did not travel beyond the hedges of my estate till after the peace of Utrecht." Some of the Essex farmers, however, kept their sheep upon turnips towards the end of the seventeenth century. Townsend, who had seen them grown as a field-crop in Hanover, made turnips and clover his great study when in 1730 he turned his back upon politics. He is said to have thereby increased the value of some of his lands tenfold. It was late in the eighteenth century before these crops were common in remote counties, such as Devonshire and Northumberland. About the same time English turnips, as they were commonly called by foreigners, began to be known in the more backward provinces of Germany. Adam Smith shows us that the change was complete by 1776, the date of his *Wealth of Nations*. He there says of turnips, carrots and cabbages that they are "things which were formerly never raised but by the spade, but are now commonly raised by the plough." (Book I. Chap. VIII.)

Gilbert White[1] has noticed the change in the food

[1] *Five hundredth points of good husbandry*, 1573.
[2] *Natural History of Selborne*, Letter 37.

of the English people and its consequences. In old days all the livestock that could be spared was killed and salted at the beginning of winter. From Martinmas to the end of Lent salt flesh, salt fish and pease pudding were the staple food of well-to-do families. Ill-cured flesh and fish, with spoilt grain, were largely consumed by the poor. The only green vegetables for winter use were grown in gardens, and were unattainable by the labourer as late as the sixteenth century. Hence the fatal prevalence of scurvy and leprosy. To this day leprosy is a frequent disease among some few communities which live much upon corrupt fish. In mediæval Europe there were lepers everywhere, and ninety-five leper-houses have been reckoned up in England alone. The last was founded at Greenside near Edinburgh as late as 1591, and the last British leper died in Shetland during the eighteenth century.

It is impossible to separate the effect of unwholesome food from the effects of bad lodging and dirty habits. The mediæval peasant lived in a narrow, ill-built hut, such as could be run up in a few hours. The floor was of earth, the roof of reeds or straw; there was no chimney, and no glazed window. Upon the ground were strewn heather or straw, which served as a lair both for the family and the livestock, for the house was undivided, and there was no other stall or pen. Soap was dear, and the peasant rarely washed. Bedding was dear, and he slept in his day clothes.

Three hundred years have wrought a great change for the better. Scurvy, leprosy, and the plague are known among us no more, and for this we have chiefly to thank our vegetables, especially the potato, the cabbage and the turnip.[1]

DUCKWEED.

July 15.—Just after the Wharfe enters Bolton Woods there is on its left bank a tract of swampy ground with ditches and pools. In summer these are overgrown with Duckweed, which is, as all the world knows, common everywhere in stagnant water. To-day I was walking to Barden when I stopped to hunt for aquatic Insects among the Duckweed. I saw a peculiar yellow light reflected from the floating Duckweed, and on looking closely perceived that almost every frond was in flower. The yellow light was reflected from the anthers, which stood out from clefts in the edges of the fronds. *Man sieht nur was man weiss.* A few years ago I had never seen Duckweed in flower, and supposed that it seldom or never flowered in England. A botanical friend, Mr. Cheesman of Selby, took me to see it in flower, and since that time I have discovered how common the flowers are and how easily they may be seen by an attentive observer.

The fronds of our commonest species (*Lemna minor*) are oval, but not quite regularly so, and biconvex or lens-shaped. One end is semicircular, and the

---

[1] The reader who desires fuller information respecting English agriculture and gardening in olden times may be recommended to study the chapters by R. E. Prothero in Traill's *Social England*, and Rev. W. Denton's *England in the Fifteenth Century*.

opposite end pointed. The two sides are seldom quite symmetrical. A ridge extends along the upper surface from the round to the pointed end, something like the ridge on a house-roof, but not nearly so sharp. From each frond a thread-like root hangs down into the water. The root ends in a

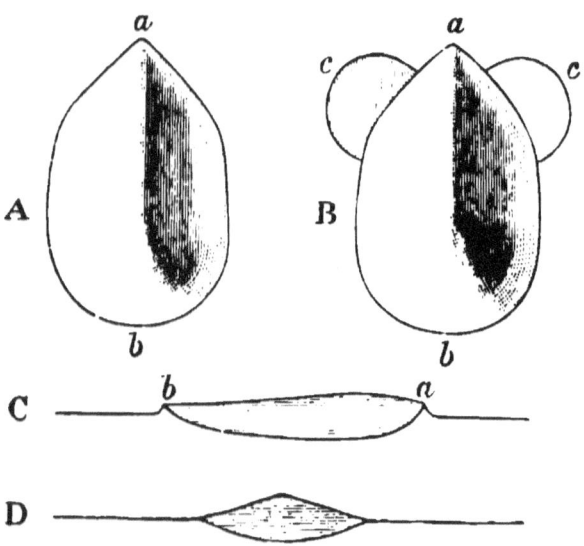

FIG. 55.—Duckweed (*Lemna minor*), magnified. A, single frond; *a*, scar of attachment to parent. A ridge extends from *a* to *b* across the upper surface of the frond, gently subsiding towards *b*; B, frond, budding out two new fronds. C, longitudinal section. D, transverse section. All the figures are diagrammatic.

root-cap, which has long been a very familiar object of study in every botanical laboratory.

We call the green discs of Duckweed fronds and not leaves, because they bear roots and flowers. Functionally they are at once leaves and stems. During the summer they bud continually. A pair of minute rudiments appears on the upper surface of a

O

frond while it is still very small and concealed within the parent frond. Each of these rudiments becomes enclosed in a special sheath formed by an overgrowth of the frond upon which it is borne. The new fronds are invariably paired at first, but one generally outstrips the other, and often only one comes to maturity. The pointed end of every frond marks the place where it was attached to its parent. Four or five successive generations may be found still fastened together, all of which are destined to break away sooner or later.

When a frond is studied microscopically, it is found to be built up of small green cells. There is a faint midrib and a pair of lateral veins. A considerable part of the interior is occupied by air-spaces, which are large in proportion to the cells and arranged in one, two, or three layers according to the depth of the frond; it is these air-spaces which give to the frond its remarkable buoyancy. The upper surface repels water strongly when the plant is in good health; the lower surface is always wet.

The flowers spring from clefts in the margins of the fronds, and are enclosed by minute scales or bracts, the outer one forming a sheath, which is burst at the time of flowering. On one or both sides of the flowering frond appears a group, enclosed by bracts and consisting of a pistil and two stamens. The pistil is flask-shaped, and surmounted by a hollow style, open at the top; each stamen bears two separate anther-lobes, slightly divided into two cells. The seeds ripen in autumn. They are minute (less than 1 mm. long), oval and ribbed along their length.

They float in water, and germinate at the surface in the course of the following spring.

Lemna minor is found in every quarter of the globe, though it is wanting, or at least undiscovered, in most parts of the tropics.[1]

Winter is naturally very destructive to the floating fronds of Duckweed. A frost kills many, and sends them to the bottom. During the milder intervals fresh fronds are budded out, but they get smaller and smaller as the light and warmth decline. These winter-fronds, which are often so small as to escape the notice of any but a close observer, are very hardy, and survive a hard frost in considerable numbers, serving, together with the seeds, to perpetuate the race. In spring they emit larger fronds, which multiply with great rapidity, and soon cover the water with a green carpet. I believe that the rapid budding-out of the new fronds is materially aided by their tendency to form strings and chains, which spread loosely and irregularly over the surface of the water. Most floating objects, such as seeds of water-plants, or bits of stick, attract one another at all points, and gather into a dense mass. But Duckweed fronds attract one another at certain points only. Hence they cling together in strings and stars with unoccupied spaces between. Some simple experiments, which can be easily set up in any household, will illustrate the principle on which the difference depends. ' Take a small cork, and cut it into a number of slices. Set these floating on water The discs of cork attract one another, and are attracted to the

[1] Hegelmaier, *Die Lemnaceen*, p. 142.

sides of the vessel. Why is this? The water rises on the sides of the discs and also on the inside of the vessel, forming in each place an *ascending capillary curve*. Descending capillary curves are also to be met with. If a lump of soot or a greased disc is set floating on water, it will be surrounded by a descending curve. Mercury in a glass dish or a barometer-tube has its free surface bordered by a descending curve. We can easily change the ascending curve of water in a glass vessel to a descending one. Take a glass or cup half-full of water; the water creeps up the side in an ascending curve. Add water until it rises level with the brim; then we say that the vessel is full. Though it is full, we can still add a considerable quantity without spilling, if we do it steadily. The vessel can be made *over-full*, when the flat surface of the water will be visibly bounded by a descending capillary curve.

We have next to remark that if two small floating objects are surrounded by capillary curves, they will either attract or repel one another according to circumstances. If the curves are alike, both ascending or both descending, the objects will attract one another; if they are unlike, one ascending and the other descending, they will repel one another. A theoretical proof of this can be given,[1] but the fact can be directly established by experiment. Take a small vessel nearly full of water, and place in it a small disc of cork or wood. If the side of the vessel and the disc are both wetted by water, there will be an ascending

[1] I have tried to put the proof in the simplest possible way in my *Object Lessons from Nature*, Part II., p. 150.

capillary curve around each. The result is that the disc will be attracted to the side of the vessel. Every time that it is moved away for a short distance it comes back again. But if we add water little by little with a syringe until the vessel becomes over-full, we convert the ascending curve into a descending one. Unlike curves are now brought together, and the disc is repelled. By sucking up a little water the ascending curve can be restored, and then the disc will be attracted by the side. Thus we may go on as long as we please, causing the disc to be attracted and repelled by turns.

The shape of one of the fronds of our commonest Duckweed (*Lemna minor*) has already been described (p. 193). At each end of the ridge which runs along its upper surface the margin of the frond is slightly raised above the water-level, and to it the water rises in an ascending capillary curve. Each of these raised parts of the margin will be a centre of attraction to a like centre on another frond. The free edge of a budding frond is also raised above the water-level, and forms another centre of attraction. Hence, when a number of fronds float upon water, they are attracted to one another at certain points, while the intervening parts of their margin come flush with the water and are inert. We can imitate the effect by models. Cut out of paper boat-shaped strips, say half an inch long, and pointed at each end. Turn every point up, and set the strips floating upon water. There will be an ascending capillary curve at each end of every strip, and these will attract one another, so that the strips will arrange themselves in chains and stars like Duckweed. They will be equally attracted by the sides of the vessel, unless this is made over-full of water.

To get direct evidence of the existence of centres of attraction around the floating fronds, set one floating upon water, and bring near it a clean glass rod dipped into the water. There will be an ascending capillary curve around the rod, and this will attract the frond, which will turn itself about, so as to bring one of its centres of attraction next to the rod. The frond may be dragged about the surface and made to turn round without being touched.

What good does the Duckweed get from these centres of attraction? It is these which cause the fronds to cohere into strings and chains instead of forming a compact mass. A moment's thought shows how profitable this is to the plant. Were the Duckweed to crowd together, like bits of cork or seeds, the fronds in the centre of the mass would be unable to get room for budding. It would be necessary to displace a vast number of mutually attractive bodies before a single new frond could be pushed out. But by the simple provision of inequalities of level along the margin, the fronds group themselves in stars and strings, with lanes between them, so that they can push forth fresh buds without difficulty as long as any unoccupied space remains.

The same capillary forces aid in the transport of Duckweed to fresh sites. If we put a stick into water overspread with Duckweed, we cannot fail to notice how the fronds cling to the stick. They cling in a particular way, which enables them to bear transport more easily. The wetted surface of the frond is attracted to the wetted stick, because both have ascending capillary curves applied to them. The water-repelling surface, which best resists drying, is turned outwards and exposed to the air. Duckweed clings to the legs of water-birds and to the elytra of water-beetles, and may be carried by them to distant pools. The wide distribution of the various species and the extraordinary speed with which they overspread any water-surface to which they may get access, are due among other things to the capillary forces which come into play at the surface of any liquid.

Routine.

The course of the year admonishes every man who takes life seriously to attend to his daily routine. Life is a long year; the year is a long day.

Here are three maxims by Lagrange, which I venture to recommend to every student. The first was borrowed from the practice of Frederick the Great.

1. Do the same things at the same hours every day, taking the hardest first, if possible.

2. Before going to sleep settle the plan of the next day's work.

3. When you read for study, read pen in hand.

To these I will make bold to add another. After the morning bath encourage the circulation by running till you are out of breath. A strong young man will easily run a mile, but those who are older or less fit should only attempt what they can do without distress.

WEEDS.

I have about an acre of ground to look after. The natural slope is so sharp that in order to get a level tennis-court and a level terrace round the house, great embankments of earth have had to be formed. Some of these are planted with evergreens ; one has been covered, with very little trouble and no cost, by the Creeping Buttercup. My gardening friends smile when I tell them that I am planting one of the commonest and most mischievous of weeds. But I am well satisfied with the plant. It forms a thick mass of green foliage, which completely hides the ground all the year round. In summer it is gay with yellow flowers. When it has once established itself no intruders can gain admission, and neither clipping nor weeding is required. But care is needed to keep the Buttercup within bounds. It is a rapid creeper, and will spread fast over ill-tended ground. If I had three or four acres to mind instead of one, I would plant no Creeping Buttercups.

The pastures which formerly occupied this site abounded with Sorrel and Earth-nut (*Bunium flexuosum*) and these are our most troublesome weeds. The tough, yellow root-stocks of the Sorrel, and the chestnut-shaped tubers of the Earth-nut enable them to offer a stout resistance to the hoe and every other weeding tool. Turn over the ground as often as you please, they come up again in undiminished numbers. There is no remedy but total extirpation one by one, a work of time and patience. Shall I ever be rid of them? Probably not, but I hope to keep them in subjection at least.

What is a weed? A plant that persists in coming up where it is not wanted. Weeds may be beautiful, at least few of us would deny beauty to the Poppy and the Dandelion and the Corn Cockle. They may even have a certain use as food or medicine. But if they invade our fields and gardens against our will, we set them down as weeds, and exterminate them as well as we can.

Provoking as they are, we cannot help admiring their cleverness. Notice the rosettes of the Rib-grass, Dandelion or Shepherd's-purse, pressed close to the ground, and denying space to any other plant within a certain radius. What an ugly bare patch is left on the lawn when one of these is rooted out! Or notice the artfulness with which many agricultural weeds time the ripening of their seeds, so that they are reaped with the corn and sown with the corn. Have they really adapted their original habits to those of the cereals, or was it only a happy coincidence? See how some weeds, like the Creeping-Buttercup, can propagate by runners, others, like Celandine, by little bulbils, small and easily detached buds, which produce new plants whenever they are cast upon suitable ground. The Speedwells and many others spring up again after they have been chopped to pieces. Agrimony, and the Forget-me-not of the fields, and Hedge Avens, and Burdock have hooked fruits, which cling to the hides of cattle and the fleeces of sheep, and so make their way into new pastures. But visible contrivance explains only a small part of the facility with which weeds spread in cultivated ground.

Canon Ellacombe in his pleasant little book on a Gloucestershire garden observes that every year there must be millions of seeds formed, and for the most part ripened, in that acre or two, yet few of them produce seedlings, while Groundsel, Thistle, and other weeds seem to have an unbounded power of germination. Yet the garden flowers have their special contrivances too, less familiar to us than those of our common weeds, partly because the exotic species are less frequent here, and partly because they are not at home with us. It often happens that an imported plant cannot bring its contrivances into action for want of a particular friendly Insect or some other favouring circumstance, which the land of adoption does not supply. Climate and soil may be adverse to imported species. Some of our garden plants come from countries which are much hotter or colder, much drier or wetter than Britain. It is to be expected that in all these matters the natives will be at an advantage in comparison with forced immigrants.

If this were all, if it were merely a question of climate and soil, or of accustomed surroundings, all plants might be expected to suffer when transported to distant continents. But when we look into the facts, we find that this is not at all universally the case. The weeds of Europe do not suffer when transported to the southern hemisphere, but flourish and often drive out the native plants. The weeds of the southern hemisphere are unable, however, to make things even by invading any patch of ground in Europe. Let us cite a few examples in proof. In Australia, such European weeds as the Bathurst Burr (*Xanthium spinosum*), the Noogoora Burr (*Xanthium strumarium*), the Spear-thistle, the Sweet-briar and the Stinging Nettle have spread far and wide, and often constitute a real plague. In New Zealand our Dock, Water-cress and Sow-thistle have multiplied so as to require proscription by law. In St. Helena the native vegetation has almost disappeared before man and the plants and animals which he has brought with him. Our common annual grass (*Poa annua*) thrives in many parts of South America, and our Shepherd's-purse, as well as our common corn-weeds, have become dispersed over almost the whole world. The most troublesome weeds of the United States are said by Asa Gray to be of British origin. It is the same with the animals. Our Rabbit and Rat and Pig and House-fly and Drone-fly seldom find a country in which they cannot multiply. The Horses of the Spanish conquerors ran wild and increased prodigiously in America.

But there is no reciprocity in the matter. Southern plants, and more rarely southern animals, do now and then get access to Europe, but they cannot maintain themselves here. All kinds of foreign plants are brought over in ballast or wool, and for a season or two they come up where they may chance to be thrown out, but when the supply ceases, native plants quickly take their place. Many an attempt has been made to establish the flowers of the Cape or South America in places of similar climate in Europe, but they have been uniform failures. The Agave (incorrectly called the Aloe) of Mexico has spread through the tropics, and has established itself on the shores of the Mediterranean, but it cannot hold its own except in spots where our common weeds refuse to grow or grow at a disadvantage. I know of not a single animal native to a distant southern country which can maintain itself in Europe.

Hooker and Wallace have spoken of the aggressiveness and colonising power of the Scandinavian flora, but this is too limited an expression. The plants, and not only the plants but the animals of Europe and the greater part of Asia exhibit this dominance; it is a feature of the Palæarctic fauna and flora. The fauna and flora of North America occasionally give way to the Palæarctic fauna and flora, but show dominance over the animals and plants of other parts of the world. What is known of the animal life of the more recent geological periods tends to show that this relation is of very long standing. Not a few animals now characteristic of distant countries, such as the Lemurs, Tapirs, Hippopotamus, Giraffe, Sea-cows, Sloths, Elephants and Marsupials, were once European. There is, I believe, nothing to show that they did not originate in the northern hemisphere. But the imperfection of the geological record bids us to be careful in drawing wide inferences.

The plants and animals of our fresh waters do not enjoy the same dominance. Our fresh waters are cut up into many small portions, and the severity of the competition in them is thereby greatly reduced. Accordingly we find that many of our fresh-water animals, especially the pond-snails, concerning which we have unusually full information, exhibit a great range in time, the genera being traced to Wealden or even to Carboniferous times. They survive by isolation, as certain ancient land-animals survive in Australia or New Zealand. It is interesting to note that our fresh-water areas can be successfully invaded. The Anacharis of North America is a familiar instance. The Azolla of North America is now plentiful in the canals of Holland, and may easily spread to other parts of Europe.

The great land-mass of the northern hemisphere for a long time past seems to have been the usual birth-place of new forms of life. Here severity of competition has created new races, which have spread into the southern lands as opportunity offered, driving before them the original inhabitants, and then themselves becoming unprogressive by reason of their isolation.

It is probable that since the time of the formation of the Chalk the great oceans have always been pretty much where they now are. There have been normally one or two continents in the northern hemisphere. When, as is now the case, North America has been cut off from Asia, the barrier has not been of a very permanent nature. An elevation of 180 feet would, as Dana remarks, form a land-passage 30 miles wide from Asia to America. The southern hemisphere is mainly occupied by sea, but possesses three continents, viz. South Africa, Australia, and South America, which have sometimes existed as islands, and have sometimes been joined to the northern continents. There is no proof that any one of the three has ever been united to another. South Africa was cut off from Europe in later Tertiary times by the sea of the Sahara. Before the elevation of the Andes, which is apparently an event of no very great antiquity, it is probable that South America was cut off from North America. The accompanying diagram shows rudely the normal arrangement of the great natural continents.

PALÆARCTIC.		
Europe.	Asia.	N. America.
S. Africa.	Australia.	S. America.

The Palæarctic region is much the largest of the natural continents, and is intersected by few insuperable barriers. Hence freedom of communication and the fiercest rivalry. Every-day experience teaches us that in any competition the severity will increase with the number of competitors, and will diminish with artificial restrictions of any kind. It is harder to be the best cricket or football player of a county than of a village, of all England than of a county. If there are scholarships to be awarded, and one is limited to natives of Cornwall, another to natives of Devonshire and so on, you will get a poorer competition than if all were open.

Dominance is much the same in the tribes of men as among plants and animals. We understand the dominance of the European better when we recollect how race after race has fought for mastery in Europe. We understand the dominance of Englishmen over remote savages better when we reflect upon the ancient wars within these islands, the "scuffling of kites and crows," when tribes of all kinds strove together with life or death as the issue. Let the survivors of such a competitive examination as that be brought face to face with some long-isolated Polynesian people, and can it be doubted for a moment which will prevail? Races of men, races of animals, races of plants, religious faiths, modes of civilisation, all originate in the northern continents, and spread out in successive waves. But there is no return-current. The plants and animals of the southern continents can no more return to Europe or Asia than the Basques and Finns can recover Central Europe. The Palæarctic Region, and in a less degree North America, have been the *officina gentium* of which Jornandes spoke, the laboratory in which new tribes are fashioned, the starting-point of waves of migration which at length reach to the remotest corners of the earth.

Our common European weeds are the very strongest in competition of all plants. They have come out first in the contest for place. Most of them produce plenty of light seeds, which are easily dispersed by the wind. Most of them are hardy and able to endure extremes of temperature. Most of them are self-fertilised, or wind-fertilised, or capable of being fertilised by any good-sized Insect, and they are therefore able to thrive in new countries, no matter what Insect-life they may find there. But our garden plants are selected by Man for their beauty. Many of them come from the southern continents, or from countries where competition is less severe than with us. Many are rare in their native land. We clear a bit of ground, plant it with a miscellaneous collection of such species, and then a struggle begins between the natives and the foreigners. It is all that we can do to keep the weeds from exterminating their feeble competitors. We tear them up by the roots, chop them to pieces with hoes, and with much pains just succeed in preserving our favourites from destruction. The naturally selected are so much stronger than the artificially selected that every three or four acres must have a man to turn the scale against nature and keep it turned.

Moorland Plants.

I wish to discuss certain peculiarities of the very commonest plants of our Yorkshire moors. It would be a great help if I could take it for granted that my readers knew the most easily ascertained facts respecting these plants. But it is only those who call themselves botanists who have attempted to study the structure of our wild flowers. The rest of the public, 99 per cent. or more of the whole number, keeps aloof and never attends to these things.

I am bound to say that the public has one excellent

reason for giving up Botany as a hopeless task, and that is, that the botanists obstinately persist in speaking a language of their own. We ought to have, as they have in Germany, descriptions of native plants in our own language, but we prefer to write our Floras in Latin and Greek. I believe that this practice is unnecessary, and further, that it is the greatest obstacle in the way of a wide-spread knowledge of the subject.

As a very young man I used to defend the learned language of Botany and Zoology, and I know pretty well the arguments that can be used in favour of it. But when I came to teach Natural History to others, I quickly felt what a hindrance the language is to those (the vast majority, of course) who read no Latin or Greek. Only a very few ever come to master it, and most of those few are the worse for what they seem to have gained. For the technical terms are allowed to count as real knowledge. The student with much labour learns to apply his rules of nomenclature to natural objects, and then thinks that he has made a step towards understanding the objects themselves. Very often he has only interposed a fresh barrier between his own mind and the world of nature. Learned words easily disguise the want of observation and thought. You may set down all the formulas respecting a plant that ingenious pedantry can devise, and yet know nothing about it that signifies. The more learned the phrase the easier it is to deceive yourself. With few exceptions every result of the study of nature which is at once well-ascertained and important can be adequately expressed in plain English ; it is only the insignificant or half understood

P

facts that call for technical expression. Elementary teaching in particular, which should be solely concerned with what is well-ascertained and important, can always be carried on in English.

I should much like to see a British Flora which would be intelligible without a dictionary to anybody

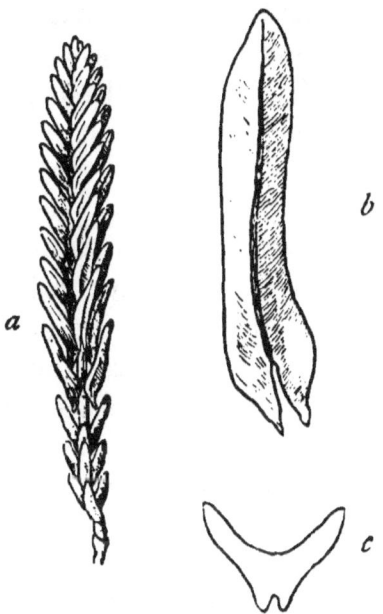

FIG. 56.—Ling (*Calluna vulgaris*). *a*, a leafy branch, magnified; *b*, a leaf, from below; *c*, cross-section of leaf.

who would first take the trouble to master the structure of half a dozen plant-types. The thing could be done, and if done in a human way, without respect to the scruples of highly special students, would do much to enlarge the body of working naturalists.

The very commonest plant of the moors is Ling,

one of the Heaths. When we speak of Heather, it is this plant which we mean. The stem is woody, tough and dry, branching continually, and never attaining a considerable height. It is densely clothed with leaves. The leaves are very small and pointed; they spring one exactly beneath another, in four rows, which run at equal distances up the stem. Each leaf is hollow above and prominent below, so that when the row closes up, as it does at certain seasons, the leaves fit neatly together. A glance at the actual plant will reveal the arrangement at once, but as the parts are minute, a lens should be employed to aid the eye. Why do the leaves overlap? I suppose because they can thus screen one another from the air. If the air is very dry or very cold, the leaves will be protected by the smallness of their exposed surface. Young and tender leaves are often protected by close packing, but in Heather the arrangement is lasting, and can be turned to account at any time.

Heather is a singularly dry plant, and for this reason it is very slow to wither. Breaking the stem across makes little difference to the leaves and flowers for a long time, for the stem transmits very little water. Heather is eminently fit to endure summer drought, which is one serious incident in the life of moorland plants, though possibly not the most trying of all.

The Crowberry, which grows so plentifully on the moors, is often taken to be a kind of Heath, and it really looks like a Heath, being of low, trailing habit, with wiry stems and crowded, evergreen leaves. The flowers, however, differ conspicuously from those of any true Heath.

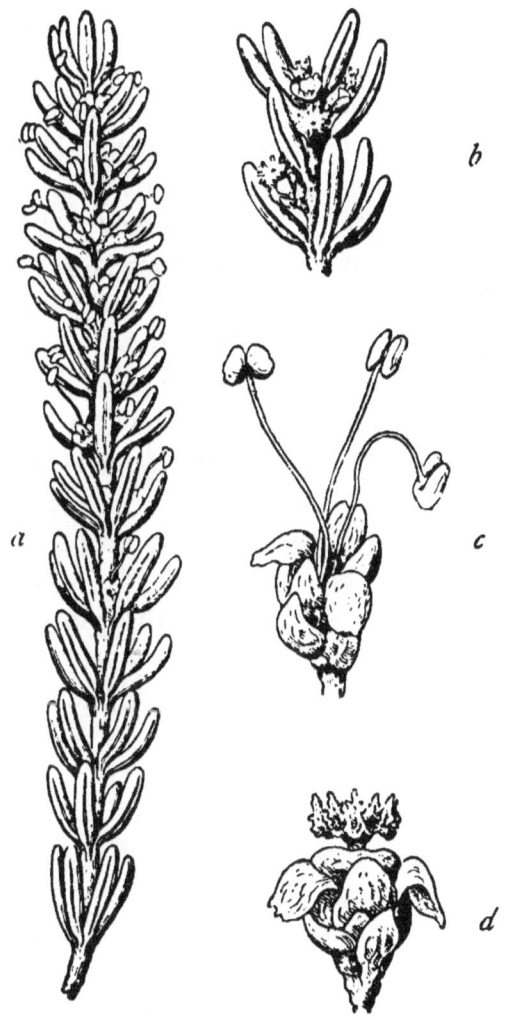

FIG. 57.—Crowberry (*Empetrum nigrum*). *a*, a staminate flowering branch, slightly enlarged; *b*, part of pistillate flowering branch; *c*, a single staminate flower; *d*, a single pistillate flower.

Crowberry flowers early in the season, often in April, and then the male flowers show their crimson stamens. So much pollen is emitted by the small but numerous flowers, that in walking over a patch of Crowberry we kick up clouds of yellow dust. Such abundance of pollen points to wind-fertilisation. Hazel, Grasses, Pines, and other wind-fertilised plants produce immense quantities of pollen, nearly all of which is wasted. Insects, if guided exactly to the right place by the formation of the flower, bring the pollen-grains surely and accurately to the stigma. Economy of this precious dust is the reward of the flowers which are able to win the co-operation of Insects. It is not every clump of Crowberry which yields pollen. About half of the plants bear blackish-purple female flowers instead of stamens. There are also a few flowers with both stamens and pistil. The pistil ripens to a small black fruit, which is harmless and even refreshing when there is nothing larger or better-flavoured to be had. Grouse devour it in large quantities, and thus sow the seeds up and down.

The leaves of Crowberry are well worth careful examination. They are smooth and glossy. At first sight we should say that they were rather succulent, but on pinching them we find that they are not succulent, only a little inflated. Along the middle of the under surface runs a white line, which we might take for a midrib. But if we examine the leaf with a lens, or better still, slice it across with a razor, we find that the blade is bent round into a hollow cylinder, and that the white line is the meeting-place of the edges. Only one surface of the leaf is exposed, that surface which in ordinary leaves is uppermost. This is smooth and shining in Crowberry, and curves completely round, ending along the white line. The proper lower surface, which bears the stomates, is rolled inwards and concealed from view. Several of our true Heaths have the leaves rolled in somewhat the same fashion, and in the Crossleaved Heath the

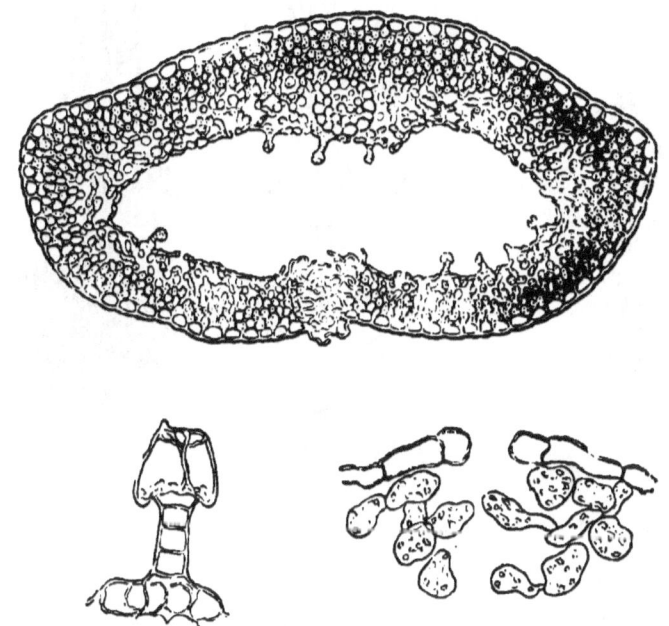

FIG. 58.—Cross-section of Leaf of Crowberry, showing the rolled-up form. The lower figures show one of the peculiar hairs and one of the stomates. Both are confined to the inner, which is properly the under-surface.

edges almost meet. *Andromeda polifolia*, the Cranberry, and certain rare British Heaths (*Phyllodoce cærulea, Loiseleuria procumbens*) also have their leaves more or less rolled backwards in the same fashion, so that this seems to be a feature common, though by no means universal, among moorland plants. Moorland plants which are not Heaths or Crowberries occasionally roll up their leaves. A certain Grass (*Nardus stricta*) which forms close tufts of fine, wiry, grey

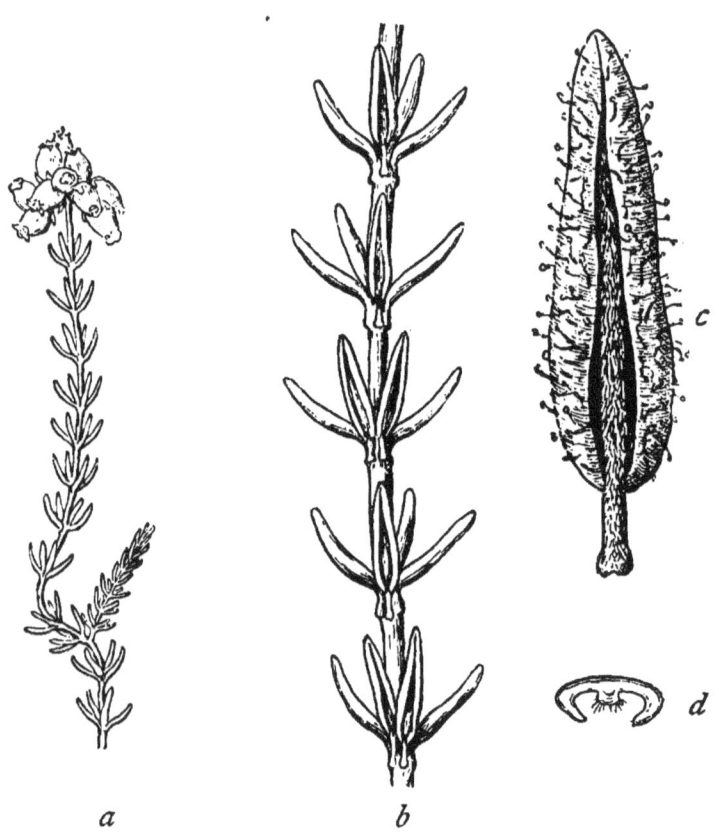

FIG. 59.—Cross-leaved Heath (*Erica tetralix*). *a*, a flowering branch; *b*, part of ditto, magnified; *c*, a leaf seen from the under-side; *d*, section of leaf.

leaves, is one of these, and its leaves look wiry or even bristle-like, because they are rolled into slender and rather stiff hollow cylinders.

This contrivance of leaf-rolling seems to be intended

to prevent the drying-up of the leaf during summer heat or parching wind. The stomates, by which water-vapour is transpired, open, not upon a free surface, but into a closed chamber, in which the air is still, being further secured against frequent change by numerous hairs, which project into it and guard the opening. The exposed surface of the leaf bears no stomates, and is overspread by a layer of cork, which renders it impervious to moisture. That Crowberry is really effectively protected against drought there can be no doubt. I gathered a plant in May, and hung it up in my study. After eight days it was still green, and capable of forming new buds when placed in damp earth. I have never known Crowberry or Nardus to perish by drought.

Kerner, in his *Natural History of Plants*, has discussed the possible uses of the rolled leaves of the Crowberry and other plants. He thinks that they keep the evaporating surface dry, and allow air to be exhaled from the leaf even when it is drenched with rain. One explanation does not absolutely exclude the other; it is conceivable that the rolled leaf may be serviceable under the opposite conditions of too wet and too dry. But I am inclined to think that the leaf would not be rolled up so completely if exclusion of water were the main object. A far less elaborate contrivance suffices in other cases to keep the under-surface of a leaf dry.

Another consideration inclines me to look upon the rolled leaf as specially a protection against drought and excessive transpiration. The in-rolled, stomate-bearing surface is often notably reduced. Striking examples could be quoted of Heaths and moorland Grasses in which it becomes insignificant in comparison with the exposed outer surface, which is rendered impervious by its dense cuticle.

Crowberry seems then to furnish a simple case of adaptation to a particular contingency of moorland life, viz., summer drought. The adaptation appears less interesting, perhaps, because it seems so obvious. We may think that we could have devised such a mechanism ourselves. But we have not yet got quite to the bottom of the question, and I fear that it will grow darker as we proceed to accumulate facts. The problems of Nature are seldom ridiculously easy.

It will occur to those botanists who have gathered Andromeda or the Cranberry that these plants are little liable to drought. Crowberry and the Cross-leaved Heath often grow on sandy slopes or among stones, where in dry weather there is no visible moisture. In summer heats there are few drier places than moorland ridges. Being high, they receive no water from rivulets, but only direct from the sky; there are no deep alluvium, no matted grasses, no overhanging trees to keep in the water of the soil. The wind blows constantly, and parches the soil still more. But the moors are not altogether dry. They are traversed by hollows, perhaps with a floor of boulder-clay or shale, and these are often choked with Sphagnum moss, which cuts off the natural outlet. In such hollows no summer heat, no east winds ever suffice to dry the soil, and these are the favourite haunts of Andromeda and the Cranberry. Yet

Andromeda and the Cranberry furnish good examples of the rolled-up leaves, which we thought appropriate to an unusually dry soil.

The common Rush (Juncus) is notoriously a native of wet places, and is often found growing in pools which never dry up. Yet it has some features in common with what are called the "Xerophilous" plants, which live in places subject to drought, and which are specially protected against undue evaporation from the leaves. The Rush has its leaves reduced to sheaths, which invest the base of the stem. The stem takes upon itself the functions of a leaf, turns green, and is provided with stomates. It is a cylinder, which of all much elongated solid figures, exposes the smallest surface in proportion to its volume; it is nearly upright, and therefore little liable to be scorched by the noon-day sun. If we were to reason from herbarium specimens only as to the habits of this plant (a most dangerous form of speculation) we might easily set down the Rush as a native of some desert tract, which had suppressed its leaves to escape perishing by drought.

The Rush, like Ling and Crowberry, has a very dry stem. The interior is filled with pith, which greatly exceeds in bulk the green layer on the outside. It would seem that moorland plants are liable to suffer from too much water as well as from too little. This is pretty certainly the case with the Rush, which grows in the wettest of the moor.

Goebel ([1]) has described for us in a very interesting

[1] *Pflanzenbiologische Schilderungen.* IV. "Die Vegetation der Venezolanischen Paramos."

paper the vegetation of the Venezuelan Paramos. The Venezuelan Andes, an eastern branch of the main chain, rise, he tells us, out of marshy, densely wooded lowlands. Starting from these, and continually ascending through the zones of vegetation described by Humboldt, the traveller at last reaches the highest tract which supports plant life. Here all is wild. The trees, crippled even at lower levels by the cold winds, are replaced by shrubs with small, leathery leaves, and these at length give up the struggle. At heights corresponding to the higher summits of the European Alps the only vegetation is scanty, dwarfish, and adapted to a rugged climate. This highest zone of plant life, which on certain peaks is cut off above by the snow-line, is the Paramos, and Goebel defines it for scientific purposes as the zone between the upper limit of trees and the snow-line. It is a region of cold winds and sudden changes of temperature. Heavy dews, and showers of rain or hail are frequent. The sun shines bright at times, but is apt to be suddenly obscured by dense fogs. Pools and bogs are common, and there is no lack of water anywhere. The great height and the consequent rarefaction of the air promote rapid evaporation, which is intensified by winds which are described as peculiarly cutting. The temperature rarely falls to freezing-point.

In such a climate, and the very description makes us shiver, Man would no doubt claim an extra greatcoat. Quadrupeds, if there are any, would be all the better for thick fur. What kind of protection can we suggest for the plants which strive to subsist under conditions so unpleasant?

The plants of the Paramos are low, an obvious advantage where the wind is cold and boisterous. Many form rosettes of leaves; some of them lay up stores of food underground; a few, and these are characteristic members of the Paramos flora, carry their rosettes on stunted pedestals (stems) which are clothed with dead leaves. The leaves of all the plants are small, sometimes reduced to needles; some are rolled-in, others woolly, others leathery; a few are pressed close against the stems which support them. We find in the vegetation of the Paramos all the characteristic features of a xerophilous flora, except that succulent plants (like Stonecrops) are few or wanting. Bleak winds, it would appear, produce much the same effect upon plants as drought.

Goebel expressly maintains this proposition, that a low temperature with wind produces the effect of drought, and cites such instances as the drying-up of our European grasses in winter, when the ground is saturated with water. In the Paramos certain plants with woolly leaves actually grow in bogs. On the Roraima Mountains (between Venezuela and British Guiana, as we used to suppose) is a small-leaved Myrtle, which might be supposed from its appearance to be adapted to a dry situation; it really lives in the spray of a waterfall, but then the water is ice-cold.

The plantations about my house testify to the scorching effect of wind. To the windward side, which here is the west, the evergreens were terribly punished by the frost of last spring, but on the leeward side many of them are quite uninjured, and have since pushed out their buds vigorously, but on this side only.

We can readily understand the drying effect of wind, especially at great elevations, but drying due to low temperature is less familiar. Sachs long ago experimented on the power of absorption of roots at various temperatures. He found that in certain plants (Tobacco) the leaves drooped when the temperature fell nearly to freezing-point, although the soil was damp enough. Turnips and Cabbages, which are naturally hardier, as natives of cold countries, endured the same temperature without visible check More extended observation has shown that plants cannot absorb water even from a wet soil unless a minimum temperature, varying according to the species, is attained. The maximum of absorption is got with a warm soil, and in certain experiments the soil is artificially warmed to promote absorption by plants. Not only is absorption checked by cold, but water already absorbed may prove superfluous and even dangerous. A cold wind or a night-frost produces most damage when the ground has been warm and absorption abundant. Some plants have a mechanism expressly devised to meet this contingency. At the tip of the leaf or leaflet are large water-pores, which discharge drops of water when sudden cold renders the quantity of water previously absorbed excessive. Many grasses and Alchemilla (Lady's Mantle) have such water-pores, and if a cold night should succeed to a warm, damp day, they exude big drops, which are confounded with dew by thoughtless people. Dew forms only beneath a cloudless sky and in minute drops. Exuded water appears whether the sky is clear or clouded, and forms large drops at a few points.

Here I will throw in a few remarks upon the watering of gardens. The gardener never waters when the sun is high, nor when a cold night is to be expected. He would waste his labour if he were to sprinkle the ground when the sun is sure to come and dry up a great part of the water almost as soon as it reaches the soil. Moreover, energetic absorption and transpiration, suddenly induced without reference to the physiological conditions of the plants, is a strong measure which often turns out ill. In such a climate as ours it is better to submit to the hardships of the changing seasons, which are seldom intolerable, rather than make violent and capricious changes in the supply of moisture. To water before a cold night may be much more serious than to water in a blaze of sunshine. The plants will be unable to absorb the water largely because of the low temperature, and what they do absorb may be distinctly injurious.

It is hardly necessary to remark that the comparative dryness of plant tissues in cold weather is the best and safest thing for them; the mere frequency of dry tissues in living plants during winter would be conclusive on this head. Dry shoots and boughs, which do not bleed when cut, will face the hardest frosts known in Europe; but in spring, when they are loaded with water and watery fluids, they will be blighted by a temperature only a few degrees below freezing-point.

My own garden has lately shown me how a cold

wind may injure plants with young leaves full of watery juices. The terrace in front of the house was not long ago covered with healthy plants sending up plenty of vigorous shoots. Three or four days of north wind came on (middle of May) and the plants were scorched and blighted. Now they look as if a sheet of flame had passed over them ; the tips of the shoots are dead, the leaves curled and blackened at their edges. No watering would have saved them, nor even diminished the injury.

Kihlman has specially noted the effect of dry, cold winds in his account of the vegetation of Russian Lapland.[1] Even marsh-plants, he tells us, perish from drought in the dry wind-storms of early spring, and the trees of Lapland are regularly cut down to the level at which they are protected by the deep snow of winter.

These considerations may well induce us to enlarge the interpretation which we first put upon the peculiar structure of the Crowberry-leaf. It is admirably protected against drought, it is true, but not against drought only. It is equally well protected against cold, and against cutting winds, which would set up a forced transpiration when the roots were unable to raise water from the dampest soil.

I am therefore inclined to look upon Crowberry, the Cross-leaved Heath, the Cranberry, Andromeda and the Rushes as needing protection against cold quite as much as against summer drought. We may expect to confirm or refute this supposition by studying the

[1] *Pflanzenbiologische Studien aus Russisch-Lappland.* Acta Soc. pro fauna et flora Fennica. Tom. VI. (1891).

distribution of these plants outside the British Isles. A little inquiry brings to light the fact that the Ling, Bilberry, Cranberry, Cowberry, Andromeda, Crowberry, together with the commonest moorland grasses, all extend far into the Arctic Circle, the Scotch Heath and the Cross-leaved Heath close up to it. The defences of which they mainly stand in need are defences against Arctic and Alpine cold, but these are effective also against the drought of an English summer. Drought and cold and wind all tend to parch the tissues or at least to cut off the supply of water taken in by the roots, and diminished transpiration is the remedy of nature for all three contingencies.

To make this clearer I will enumerate the defences of desert plants against extreme drought, and append to each a short list of Arctic, or at least high northern plants which exhibit the same feature.

Xerophilous or desert plants often exhibit one or more of the following features:—

1. The leaves are reduced, rudimentary, or wanting altogether. Among Arctic or high northern plants various species of Juncus (Rush) yield examples of the same thing. Furze (which, however, is not Arctic) has trefoil leaves as a seedling, which are afterwards replaced by spines, which contain chlorophyll and possess stomates. Pines, Juniper and many Heaths have the leaves much reduced in size.

2. The leaves are rolled up. Crowberry, etc., are Arctic examples.

3. The leaves are closely imbricate, or pressed against the stem, or pressed against the ground, in the last case usually forming rosettes. All these arrangements

diminish the free surface of the leaves. Heather, many Saxifrages and Sedums are northern forms belonging to this class.

4. The leaves are woolly. The Edelweiss of Switzerland is an Alpine example of the same thing.

5. The leaves are succulent. The Sedums of our northern highlands exhibit the same adaptation.

Goebel has pointed out that maritime and saline plants often exhibit contrivances which check transpiration, even in plants which are abundantly supplied with water. Samphire is possibly a case in point, but the rocky situations in which it grows are arid as well as maritime, and it is not certain that its peculiarities depend upon its being wetted occasionally with salt spray. The Glass-worts are a better example, as they grow on muddy sea-shores. They have the leaves suppressed and the stems fleshy, like so many plants of the desert. The little Frankenia, found in salt marshes on our south-east coast, has rolled leaves, like Crowberry. Thickened epidermis, woolly leaves, and concealed stomates are also to be found among the plants of the sea-shore. Here, though water is plentiful, it can only be procured by separating it from salts which greedily absorb water and do not readily part with it.

The one point common to maritime and Alpine plants, I mean the difficulty which they have in absorbing water, however abundant the supply, may possibly have something to do with a fact, long known and often speculated upon, viz. that certain species are found in both situations. Scurvy-grass, Sea-thrift, Sea Plantain and Sea Pearlwort are examples. We find all these among the Yorkshire hills and on the sea-coast, but-in no intermediate places. The plant which has been acclimatised to one of these habitats is thereby adapted to face the most serious difficulties of the other.

Hindered transpiration points to a limited supply of water, which may arise in various ways. (1) Water may be deficient altogether, as in the desert, or on rocks and loose stones. (2) Water may be plentiful, but absorbed with difficulty because of low temperature. (3) Water may be plentiful, but absorbed with difficulty because of salts dissolved in it. The Crowberry and our other native moorland plants are efficiently protected against contingencies 1 and 2, of which the second is probably the more frequent and serious.

A considerable proportion of our moorland plants (Ling, Crowberry, Nardus, Juncus, Cranberry, etc.) are evergreen. In this the moorland flora resembles that of high northern regions. I am not quite certain why a large part of the more conspicuous plants of cold regions should be evergreen. The explanation may be that even during the summer the conditions require checks to transpiration, and that these checks, once developed, enable the plant to endure winter conditions without loss of leaves. The glossy cuticle, the simple form, the reduced size, the dry texture, and the concealed stomates, which are common features of leaves exposed to sun and wind, form a great part of the special equipment of an evergreen.

But there are puzzling exceptions to the evergreen habit. Take the British Heaths (*Ericaceæ*) for example. Of the genus Vaccinium two species, the Cranberry and the Cowberry, are evergreen, but the Bilberry[1] and *V. uliginosum* are not ; Arbutus is evergreen ; one species of Bearberry has evergreen, the other deciduous leaves ; Andromeda, Erica (five species), Calluna (Ling), Dabeocia, Phyllodoce, Loiseleuria and the five species of Pyrola are all evergreen.[2]

As to the Bilberry, which is one of the few deciduous Ericaceæ, it is to be remarked that it is not a characteristic moorland plant, but overspreads the grassy borders of the moors, where it chiefly competes with Ling and the grass Nardus. In spring and early summer the quick growth of the Bilberry shoots enables them to overtop its rivals, and thus to get an advantage which lasts all through the season. When winter approaches, the Bilberry gives up the struggle, casts its leaves, and appears to die down. It does not really do so, however. The younger stems remain green, and as they are provided with numerous stomates, they no doubt assimilate during the short hours of winter sunshine. In summer the deciduous leaves of Bilberry can safely expand far more freely

---

[1] The leaves of Bilberry often remain green through the winter in sheltered places, but elsewhere they are usually deciduous.

[2] The word *evergreen* does not always bear precisely the same meaning. It is applied to leaves which are able to endure frost, and last through a great part or the whole of winter, being all renewed at once in spring. Most of the Ericaceæ are only evergreen in this sense. In other cases the leaves are changed a few at a time, usually in summer. The tree is never bare, and the leaves may last more than one year.

than those of its evergreen rivals, which are necessarily minute, and this may possibly be a consideration of weight.

The leafy shoots of Ling and Bilberry, which seem to spring out of crannies between rocks, are sometimes borne upon long woody stems, which have made their way up from a considerable distance. In rocky places what look like tufts of low shrubs are sometimes the tops of small trees. I have traced some of the ancient, woody stems for many feet among the loose stones.

It is, I think, worth while to attend to any peculiar features of particular plants and animals, and to interpret them as well as we can. But our interpretations are never complete. We see some way into the problem, and then are baffled by our ignorance and by the complexity of the case. We can rarely apply the *experimentum crucis*, or find decisive instances. There is always, or nearly always, as Goebel says, some unknown quantity which decides why of two plants similarly situated, one will show conspicuous adaptations to its surroundings, while the other will not. For this reason the methods of biological inquiry are apt to be loose in comparison with the methods of the physical sciences. In order to prove that the height of the mercury in the barometer depends upon the pressure of the air, we try to show that among varying conditions of moisture, temperature, light and so forth, the height of the mercury goes up or down as one of these conditions, viz., the atmospheric pressure, increases or diminishes. If it appeared that the mercury was largely influenced by heat, or that while some mercurial barometers rose and fell according to pressure, others of rather different construction rose and fell according to the intensity of the sunlight, the present doctrine of barometric changes would be destroyed. The barometer is, so to speak, actuated by a single string. But plants and animals are things of complex behaviour; they are actuated by many strings, and we never know when we have found them all out. This is, I imagine, why what I have called "negative exceptions" prove little or nothing in Biology. You find that pulling a particular string produces a certain action upon A. But B has no such string, though you can give no good reason why it has not. Rare indeed are the cases in which we can reason out a direct test, by which our biological speculations are to stand or fall. Complexity, unfathomable complexity, on the part of Nature, and ignorance on our side, preclude decisive experiments. But in a humble way we may observe, and speculate, and try. We shall not get the certainty of physical demonstration, but we may hope in time to become reasonably sure of interpretations more directly interesting to mankind than any other conclusions of Science—more interesting because they bear so immediately upon the great and mysterious problems of Life.

THE LOVE OF MOUNTAINS.

August 1.—Simon's Seat, between Barden and Pately Bridge, is one of the chief hills in this part of the country, rising to near 1,600 feet above sea-level. I have been to the top, to look once more at the saddle-like arrangement of the rocks, discovered by Mr. Dakyns of the Geological Survey, and described with figures in Prof. Green's *Physical Geology*. The ascent is a delightful one, wood and water alternating with moor and grassy slopes. The way leads through Bolton Woods and the Valley of Desolation. There is considerable variety in the composition of the rocks, which means variety in the vegetation and variety in the animal life. I know of no better botanising ground. Insects and Birds are plentiful, and the very streams are full of life.

To take pleasure in such a ramble up-hill, for we can hardly call it mountaineering, is, I believe, a discovery of modern times. Xenophon writes as if he enjoyed a hunt upon Mount Pholoe. He must have appreciated the exhilaration which springs from active exercise in the open air. Hunting on foot he praises as good for the health, the eyesight and the hearing. He thinks it an excellent way of keeping off old age, and training the body for the hardships of war. But he gives no hint that he ever went out on a hillside without dog and net. It is the hollows, the plains, the woods and the rivers, which Virgil chiefly loves. The Alps struck the ancients with horror though they delighted in the soft scenery of the Italian lakes.

Perhaps the first man who ever climbed a mountain in order to gaze from the top, and then wrote an account of what he had seen, was Petrarch. Living at Vaucluse, near Avignon, curiosity moved him and his brother to ascend the Mont Ventoux, a low Alp, under 7,000 feet. They were warned by an old

shepherd that no one had been up the mountain for
fifty years, and that nothing was to be seen upon it
but rocks and brambles. Still they persevered, and
at length stood on the summit. The view was
superb. The Alps seemed close at hand ; the sea,
the valley of the Rhone and the mountains about
Lyons were in full view. Petrarch's thoughts ran
much upon the mountains famous in literature, upon
Olympus and Athos, and Hannibal's passage of the
Alps. After a time he took out of his pocket a
volume of St. Augustine, and lit upon words which
rebuked those who wonder at the mountains, the sea,
and the stars, but neglect themselves. He descended
in silence, reflecting that there is nothing admirable
except the mind.

Gray's diary of a tour in the North of England,
though written as late as 1769, is among the earlier
indications of interest in wild scenery. Till then the
hills had been despised for their barrenness, and
dreaded for their ruggedness and danger. It was
only when better police and better roads had driven
away fear that men began to make mountains their
playground.

Rousseau's *Nouvelle Héloïse* (1760) may be said to
have first awakened a lively interest in Swiss scenery.
Within forty years of the publication of that novel
more than sixty descriptions of travels in Switzerland
appeared. Gibbon in 1785 was astonished at the
crowds of English who haunted the lake of Geneva.
Goethe and Byron drew inspiration direct from
Rousseau, and later writers, who perhaps never read
Rousseau attentively, such as Renan and Ruskin,

exhibit that sentiment of Nature which was hardly known until Rousseau's writings had pervaded Europe.

Our English Lakes began to be overrun by tourists late in the last century, as we learn from Scott[1] and Wilberforce.[2] At first the visitors seem to have kept mostly to the lower and safer ground ; the narrative which follows marks the close of the unadventurous age. By 1805 almost all the summits of the Lake hills had become familiar to thousands of active Englishmen.

In the *History of Cumberland*, by William Hutchinson,[3] a highly-coloured description is given of an ascent of Saddleback. Those who know the mountain only on its Keswick side should understand that to the S. and E. it is much more abrupt. Sharp Edge might even be dangerous to an inexperienced climber with a weak head. The summit is only 2,850 feet above sea-level. The description follows :—

"A friend has indulged us with the following description of his view of Saddleback, and the curious crater and lake there, where the lava of a vulcano is unquestionably to be found in large quantities. His tour was made in 1793.

" He speaks with great respect, in the first instance, of one Mr. John Graves, who gave him the earliest description of those scenes, and excited his curiosity to visit them ; and of Mr. Thomas Clement, a resident of the skirts of the mountain, who attended him and

[1] *Guy Mannering*, Chap. XVI.
[2] *Life*, Vol. I., p. 183.
[3] 2 vols., Carlisle, 1794. See Vol. I., p. 423.

his party on the view. . . . He says Mr. Clement lives about a mile and a half eastward of Threlkeld, at the foot of the mountain, from whose house the party proceeded about one o'clock, p.m.—That they made their passage in an oblique direction up that part which is called *Scales-fell:* and he proceeds in his description thus:—' When we had ascended about a mile, one of the party, on looking round, was so astonished with the different appearance of objects in the valley, so far beneath us, that he declined proceeding. We had not gone much further, till the other companion (of the relator) was suddenly taken ill, and wished to loose blood, and return. I was almost ready (adds he) to give up my project, which I should have done with great reluctance, as the day was remarkably favourable, and exhibited every scene to the greatest advantage.—Mr. Clement assured us, if we proceeded a little way, we should find a resting place, where the second defaulter of our party might recover the effects of his journey. After labouring another half hour, we gained the margin of an immense cavity in the side of the mountain, the bottom of which formed a wide bason, and was filled with water, that from our station looked black, though smooth as glass, covering the space of several acres. It is said to be so deep, that the sun never shines upon it, and that the reflection of the stars may be seen therein at noonday; but that was a curiosity we did not enjoy. From our station there was a gentle declivity to a smooth and verdant lawn, several yards in breadth, which was the situation our guide had promised us; and the descent thereto led us about half way to the lake: a like easy descent would have led us to the edge of the lake, round which there appeared a broad green walk; but our leader informing us of the danger of passing that slippery path, we did not proceed. We now contemplated the scene with *awstruck-wonder*. We stood directly facing the middle of the mountain, the form of which gives it the name of SADDLEBACK: and to the lake, a perpendicular rocky precipice presented itself, extending to the north-east side of the mountain, called *Foul-cragg*. To the right hand, the steepness of the rocks gradually declined; above us, and on the left, they were stupendous and perpendicular; so that in one half of the circle the rocks were lofty and precipitous, whilst in the other half they gradually decreased. My fellow traveller would proceed no further, and with my guide I was left to explore the other parts of the mountain. Winding round, and keeping the cavity on our right, we attained the ridge or summit of the rock, where we found a passage three or four yards broad: on the right, the descent to the lake looked truly awful, whilst the steep rocks on the other side were lofty, and not to be climbed by human steps. This passage, some hundred yards in length, may be compared to a bridge covered with grass. Having reached the summit, we went to the point nearest to Keswick vale, and there enjoyed a most delightful prospect; from thence we passed to the next point, being *Foul-cragg*, with Skiddaw on the left; from whence we looked down into a dreadful abyss, the bottom of which the eye could not penetrate: sheep frequently perish in this place, as the number of dead carcasses and skeletons evinced. We walked back by the side next to the lake, but to look down from thence was so terrible, I could not endure it a moment. We perceived from thence, that my companion, whom we had last left, was laid upon the ground; I pressed the guide to hasten to him, but he refused, alledging that a fog was rising, and it would be very hazardous for me to explore my way alone down the mountain: in a short time we were enveloped in a very dense vapour, so that we were obliged to keep near to each other; the sudden change was almost incredible. It was with difficulty my guide regained the passage, or dry-bridge, which we missed on several attempts; and one incautious step would have plunged us in the horrid abyss. The fog soon afterwards dispersed, as precipitately as it came on; and left us again under a serene sky. We passed to the foot of *Foul-cragg*, to view its wonderful precipices from their base; and again safely reached Mr. Clement's house, after a laborious travel of four hours.'"

A plate in the *History* shows Skiddaw as a volcanic crater, while Saddleback is a hill which positively overhangs on one side. So greatly did the Lake hills impress the imaginations of that generation! Green and Otley, in their ascent of Sharp Edge, found themselves "reduced to the necessity either of bestriding the ridge or of moving on one of its sides, with our hands lying over the top, as a security against falling into the tarn on the left, or into a frightful gully on the right, both of immense depth."

Years ago I used to read these old narratives with unqualified amusement. But a solemn feeling now mingles with my recollections of the innocent little peaks of our familiar Lake-country. On Dec. 31, 1893, in the height of his manly strength, and in the full enjoyment of his great gifts as an investigator and a teacher, Arthur Milnes Marshall fell from near the summit of Scawfell, and perished in a moment.

The Reversed Spiral.

The tendrils of the Red Bryony in our hedges or of the Passion-flower in greenhouses have been often admired by the readers of Darwin. When our eyes have been opened by the penetrating observations of the great naturalist, it is easy to appreciate the effectiveness of the slight but powerful attachments by which the tendril-bearer draws itself up to its support. An unattached tendril, when it contracts, forms a spiral running in one direction from base to apex. But a tendril which has grasped a fixed object becomes wound from right to left in one part of its length, and from left to right in the remainder, a short straight portion uniting the two spirals (fig. 60). Darwin points out that the spiral spring gives great elasticity to the tendril, a valuable quality in stormy weather. The Bryony rides out the gale with a long range of cable paid out. But why is the spiral reversed? Darwin explains that every turn in the spiral twists the tendril once. Thirty turns in the same direction would twist the tendril thirty times in succession. No tendril of ordinary length could stand so much twisting; it would inevitably snap across.

But every turn in the opposite direction takes off twist, and a tendril twisted fifteen times from right to left and fifteen times from left to right is not strained at all.

There is another advantage in the reversed spiral;

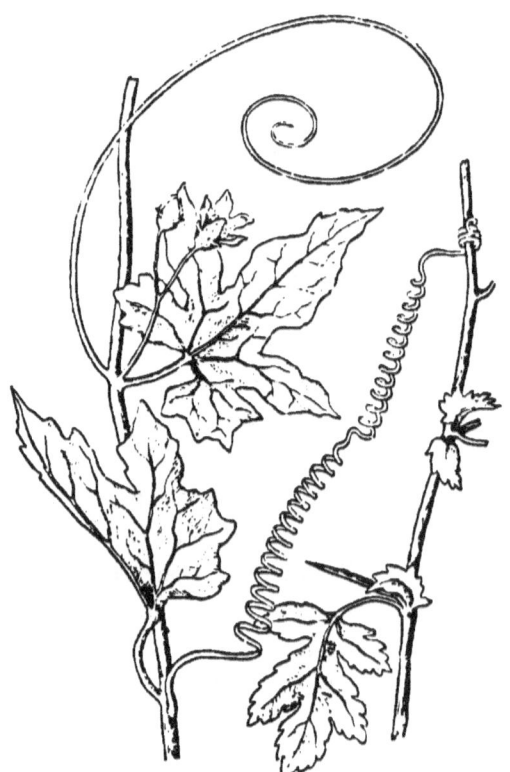

FIG. 60.—Red Bryony, with two tendrils, one free, the other clinging to a branch of Hawthorn.

it can be pulled out straight without a kink. Pull at the ends of a continuous spiral, and the turns cannot be effaced, though they may be reduced to as many twists.

The reversed spiral seems therefore to be one of the most perfect contrivances in Nature. I think I see Darwin's admiration of it in many of his expressions.[1] Nevertheless Darwin, like Mohl before him and Sachs after him, was aware that the reversal of the spiral is a mechanical necessity. When a band whose ends are not permitted to revolve has a tendency to curl, and consequently to form spirals, it can only form a reversed spiral. This is most easily demonstrated by taking a long ribbon of sheet brass, and winding it

Fig. 61.—Strip of sheet brass, which has wound itself into a reversed spiral.

into a close coil, as one would wind a tape measure. If one end of such a coil is pulled out, it will take the form of a continuous spiral. Unroll the entire coil, and hold out the ribbon straight and flat. Then gradually relax the pull. The tendency to coil will throw the ribbon into spiral turns, the middle point will revolve, and for every turn from right to left a turn from left to right will appear. The reversed spiral is not a contrivance at all; it is a mechanical necessity when a band whose ends are not free to revolve is thrown into coils.

The reversed spiral is often found in Nature where an elastic spring would be quite unnecessary. It is

[1] *Climbing Plants*, Chap. IV.

simply a convenient way of coiling a tube which has greatly increased in length while its ends were fixed. The intestine of the Tadpole is at first straight. Afterwards it grows very long, to suit the vegetarian diet of the young animal. This long tube must needs be coiled, for the space into which it has to be crowded is small. Being fixed at both ends it cannot be coiled continuously in one direction. The watch-spring coil, which represents the intestine of the Tadpole in many standard books, is a mechanical impossibility.

The colon of a Ruminant is extremely long, and having lengthened while the ends were fixed, it is coiled in a reversed spiral. It lies nearly in one plane, and winds inwards in a regular spiral to the centre; then reverses its course, and winds outwards between its former turns. The pattern is very characteristic, and immediately recognised a second time. I remember seeing it depicted over and over again in Italian pictures of the torments of the damned. At the Campo Santo at Pisa and elsewhere, the entrails practised upon by demons are shown with this Ruminant

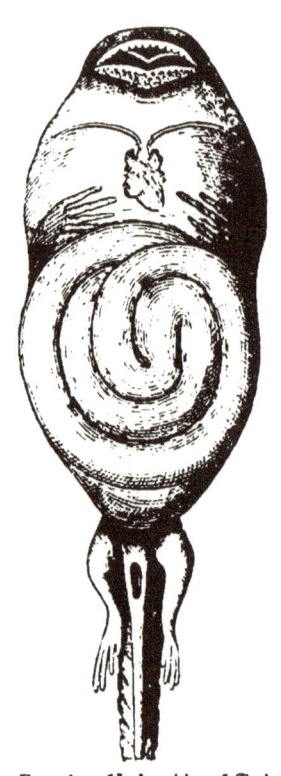

FIG. 62.—Under-side of Tadpole showing the intestine with its reversed spiral through the transparent body-wall. From Howes' *Atlas of Biology*.

feature. The kids, which an Italian butcher hangs up at his door with all the viscera exposed, soon made it clear where the old painters, careless of comparative anatomy, had got their reversed spirals. The intestine of a Pond Mussel, the vasa deferentia of Crustaceans and many other organs of various animals furnish examples of the same thing.

Gossamer.

Sept. 9, 1894.—A calm, bright autumn day. At sunrise the sky was clear, and the air perfectly still. The sun shone uninterruptedly through most of the morning, but in the afternoon, haze and thin clouds were prevalent. During the day, light and variable winds alternated with perfect calm. The evening was clear, with a gentle northerly breeze.

At breakfast-time this morning the lawn was covered with dewy cobwebs, and on walking out a little later, they were found to overspread the fields, lanes and thickets. Most of the webs were deserted, but here and there dead or living Insects—chiefly Aphides and small flies—were entangled in them. There were also very many threads attached at both ends, traces probably of the ordinary excursions of a variety of Spiders, rendered unusually distinct by the fine globules of dew. Small and apparently immature Spiders were occasionally seen, in one or two cases in little companies. They paid no attention to the captured Insects, but travelled along the webs. As the sun got higher, the Spiders became more numerous, as if they had crept from their retreats, and they were more active than before. About an hour before noon, the day being now warm and the thermometer at 62°, very many Spiders were running busily to and fro. They were chiefly bent upon mounting some elevated object, and hardly any tall weed, straw, gate-post or boulder could be examined without finding one or more Spiders climbing upon it. They were not all of one size, nor all of one species. Though very slightly acquainted with the classification of Spiders, I think I may venture to say that at least three species were represented. There were a few large and probably old Spiders in the throng, but these took no part in the manœuvres next to be described.

Having gained a post of vantage, every little Spider reared itself upon its legs, and emitted one or more threads. I could not decide whether the threads ever emerged separately from the body, as this could only be ascertained by a very close examination with a powerful lens, and the Spiders were too timid and wary to allow of this. In some cases, the threads merely crossed one another, and adhered. Attus is said to emit a brush of threads at once. The light threads, so fine that they were completely invisible when dry and single, except where the sunlight was reflected from their shining surfaces, seemed to catch the gentlest puff of air, and were for the most part extended horizontally. When the little Spider felt the pull of the threads, she let go, and was wafted along. I saw some steadily ascend, while others fell gently to the ground. In the course of

R

about an hour, perhaps twenty Spiders were observed to ascend.

By noon, the number of Spiders on the ground had sensibly diminished. In the afternoon, the sky being now to some extent overcast, many strips and patches of web were seen to descend. Though very numerous, they were by no means conspicuous, owing to the want of bright sunshine. I remember to have seen many years ago, descending flakes of gossamer glistening in the sunlight; but this time their appearance was not striking. Any person who walked along intent upon business would have neither seen nor felt anything of the gossamer.

Next day something of the same kind was observed ; but the numbers of the Spiders were greatly diminished. Several fine days followed, and very likely the Spiders congregated and took flight again. Unfortunately, I was too much occupied to attend to them ; if much is to be seen, the day must be given up to observing.

On reading what Gilbert White, Blackwall and other naturalists have observed about gossamer, I find that the Spiders which rise in the air belong to many species and genera. They rise only in still, bright weather, and gossamer in the air is always preceded by gossamer on the ground. September and October are the months in which it is most commonly seen. The Spiders often float to a great height, several hundred feet at least. The flight cannot be directed or regulated by the Spider. Spiders never voluntarily ascend upon webs, but only on fresh-spun lines. It is unusual, perhaps unexampled, for good-sized Spiders to ascend, and all that I have seen were so small as to be invisible without close attention, the body being about an eighth of an inch long or less; some of the floating Spiders, however, though of small size, are believed to be adult.

The question has been raised whether the Gossamer Spiders can, strictly speaking, emit threads from their spinnerets, or whether wind is necessary to draw out the threads. Blackwall[1] gives us the interesting results of his experiments. " Having procured a small branched twig, I fixed it upright in an earthen vessel containing water, its base being immersed in the liquid, and upon it I placed several of the Spiders which produce gossamer. Whenever the Insects thus circumstanced were exposed to a current of air, either naturally or artificially produced, they directly turned the thorax towards the quarter whence it came, even when it was so slight as scarcely to be perceptible, and elevating the abdomen, they emitted from their spinners a small portion of glutinous matter, which was instantly carried out in a line, consisting of four finer ones with a velocity equal or nearly so, to that with which the air moved, as was apparent from observations made on the motion of detached lines similarly exposed. The Spiders, in the next place, carefully ascertained whether their lines had become firmly attached to any object or not, by pulling at them with their first pair of legs; and if the result was satisfactory, after tightening them sufficiently they made them fast to the twig; then discharging from their spinners, which they applied to the spot where

[1] *Linn. Trans.*, Vol. XV., p. 455 (1827).

they stood, a little more of their liquid gum, and committing themselves to these bridges of their own constructing, they passed over them in safety, drawing a second line after them as a security in case the first gave way, and so effected their escape. Such was invariably the result when the Spiders were placed where the air was liable to be sensibly agitated: I resolved therefore to put a bell-glass over them; and in this situation they remained seventeen days, evidently unable to produce a single line by which they could quit the branch they occupied, without encountering the water at its base; though on the removal of the glass they regained their liberty with as much celerity as in the instances already recorded. This experiment, which, from a want of due precaution in its management has misled so many distinguished naturalists, I have tried with several of the Geometric Spiders, and always with the same success. Placed under the bell-glass, or in any close vessel, they in vain endeavoured to make their escape from the branch to which they were confined; but in the disturbed air of an inhabited room, they readily accomplished their object."

The rising of gossamer has been attributed to the low specific gravity of the Spider and to imaginary causes which will bear no investigation. Since Blackwall's researches it has been agreed that light currents of air are sufficient to explain the rise of fine filaments. It is not a question of buoyancy, but of surface in proportion to weight. The webs rise for the same reason that fine dust rises in moving air, and fine sediment in moving water. It is for the same

GOSSAMER

reason that impalpable drops of water form mists and clouds, instead of sinking at once to the earth. The smaller the particles, the greater the ratio of surface to volume, the greater the ease of transport by a current, and the greater the resistance to falling through air or water. If we divide a sphere into spheres of $\frac{1}{2}$, $\frac{1}{3}$, $\frac{1}{4}$, etc., the original diameter, the aggregate surfaces increase as 2, 3, 4, etc. The same is true of a cylinder or any other figure, provided that the parts are similar in shape to the original figure. It is a general and well-known law that the surfaces of similar figures increase as the square, but the volumes or weights as the cube of any linear dimension. Hence, if the scale is enlarged, the weight gains upon the surface; if reduced, the surface gains upon the weight. Take a knitting needle $\frac{1}{16}$ inch in diameter. Reduce it in all its dimensions until the diameter becomes $\frac{1}{10000}$ inch, which is of the order of fineness of a thread of gossamer. The surface is reduced to $(\frac{16}{10000})^2$, the weight to $(\frac{16}{10000})^3$, that is, the surface gains on the weight about 625 times. The actual knitting-needle falls quickly through the air, while the reduced knitting-needle, if we could make such a thing, would fall slowly, for the resistance of the air in proportion to the weight has been increased 625 times. Not only do the lightest breezes set up by differences of atmospheric pressure suffice to waft the Spiders, but they ascend when there seems to be no movement in the air, except the ascending current due to the heated ground. Such currents will cause soap-bubbles to ascend, which will not rise in-doors.

Incomplete information prevents us from clearing

up the relation of these aërial excursions to the life-history of the Spider. The circumstance that it is largely though by no means exclusively immature Spiders which take to flight connects this case with the larval dispersal of very many marine animals. The heavy-armoured adults which haunt our shallow seas are obliged to keep near the same spot, and dispersal is effected by the fresh-hatched larvæ, which often migrate before they have acquired a mouth or stomach, and are provided with temporary locomotive organs for this very purpose. In the case of land animals, where the weight of the body cannot be supported by a dense medium, locomotion is too difficult to be effected by very immature individuals, and only full-grown animals migrate; (Insects, Frogs and Birds furnish plenty of examples) but for the peculiar flight of Spiders small size is essential, and this one circumstance may have determined their deviation from the common practice of land-animals.

Spiders often protect their eggs by cocoons, which may be laid in crevices or webs, or carried about by the female. The fresh-hatched young often creep about within such a cocoon or web for some days, during which time they are watched over by their mother. At last they begin to seek their own food, which they procure by hunting. Probably no very young Spider is able to make a snare. By the end of summer, when food begins to be scarce, the young Spiders set about the business of dispersal. It is not likely that they get much to eat until the following summer, but this is a point on which we have few or no observations. We may suppose (until the point is cleared up), that they retreat to hiding-places, and like older Spiders, endure long abstinence with impunity, procuring chance supplies of food at long intervals. The power of emitting silken threads is commonly used in the excursions of the young Spider to enable it to climb from twig to twig, before it is turned to account in aërial voyages.

So long an interval separates the hatching-out of the Spider from the time at which it begins to make a web of its own that it may be given as a pure case of constructive instinct. There is no parent to show it how webs are made, nor can it be supposed to remember the minute details of the web in which it may possibly have been reared. How little we know, or rather, how entirely ignorant we are of the means by which the practical experience of by-gone generations is handed down to animals which have no occasion to apply it until they have long been separated from their own parents!

Dr. Lincecum [1] tells us that the mother and young of the Gossamer Spider of Texas ascend together. Probably this is a species of small size.

Darwin's account of the South American Gossamer Spiders is well worth reading.[2] When the *Beagle* was sixty miles distant from the shore, vast numbers of small Spiders settled on the ship. A steady though light breeze was blowing off-shore. Each Spider was seated on a single thread. All were of one species, but of both sexes, together with young ones. Another

---

[1] See *Amer. Nat.*, Vol. VIII., p. 593 (1874).
[2] *Naturalist's Voyage*, Chap. VIII.

South American species, observed on shore, darted forth four or five threads from its spinnerets, which were more than a yard long, and diverged in an ascending direction. The Spider then suddenly let go its hold, and was quickly borne out of sight.

FLOWER-HAUNTING INSECTS.

Sept. 29, 1895.—We have had a glorious September, hot and sunny. But for one thunder-shower there has been no rain, and for several days past the thermometer has regularly exceeded 80° F.

I have been noticing with some care the Insects which haunt the clumps of Asters in the garden. There are Bees of at least seven different species, Wasps, two small Beetles in scanty numbers, and a host of Flies. The Red Admiral and Small Tortoise-shell Butterflies flutter about continually, but pay no special attention to the Asters. Among the Flies I have identified two common species of Eristalis, a Volucella, a Syrphus, a Dung-fly, the metallic-coloured Lucilia Cæsar, the Blow-fly (Calliphora), and there were also several small Muscidæ which I did not examine.

There were many Insects in the air, chiefly Diptera, which did not alight on the flowers. I was able to recognise small swarms of three species by peculiarities of hovering. As I was sitting this afternoon in the sun with a book on my knee, small Gossamer Spiders now and then descended from the air upon the page. I have not seen any gossamer on the ground for several days. The little Psychodidæ abound on the window-pane.

I have made it my business for some years to hunt out the larvæ of our common Insects. I have searched the waters, both stagnant and flowing, and have pried into all accumulations of decaying organic matter that I have come across. I have particularly attended to the early stages of the Diptera. But I have to confess that nineteen-twentieths of the Diptera now buzzing about in my garden are known to me, if at all, only as items in a catalogue. No doubt a large proportion have been reared close at hand. But they are so well hidden, and the naturalist is so blind, that it is only when he sees the swarms of winged Insects that he becomes conscious of the multitude of larvæ and pupæ which he has overlooked.

It is interesting to note that Insects of very different kinds haunt flowers for honey or pollen. The Insects just enumerated pass the larval stage in various situations. Some feed on green leaves, some on decaying animal matter, one haunts the nests of Humble-bees as a parasite, some live in stagnant pools. But though they are so widely separated during the feeding-stage, the quest of honey brings them together, as soon as they have got their wings.

The honey-sucking Insects are mainly Lepidoptera, Bees and Diptera. With unimportant exceptions, all Lepidoptera, which feed at all, visit flowers. Bees make the greatest use of honey and pollen, feeding upon it in all stages. They possess the most elaborate collecting apparatus, and it is the Bees which have acted most powerfully upon the organisation of flowers. The honey-sucking Flies are few in number, but of considerable practical importance. The form, colour, and scent of some orders of flowers have been distinctly modified in consequence of their visits. As a rule the Flies have a short proboscis and prefer open flowers, but some, like Eristalis, have a long proboscis and can explore tubular flowers, as we have already seen. Their taste in colour leads them to prefer pale, dull, or speckled flowers, and their favourite odours are unpleasant to man. Bees and Moths come nearer to ourselves in their preferences, both as to colour and scent.

It is a striking proof of the importance of Insects in nature that they should have been able to call into existence a profusion of beautiful flowers. All the flowers of the garden and conservatory are in a sense the work of Insects. What they found ready to hand was a multitude of green or dull-coloured flowers of small size, without honey or scent; their visits have done all the rest.

Flowers have done as much for Insects as Insects have done for flowers. Flowers are to innumerable tribes of Insects all that domestic animals and cultivated plants are to mankind. Honey, which may be considered a joint product of the flower and the Insect, owes its great value to three properties. It is fluid, it is highly nutritious, and it can be stored without undergoing putrefaction. Its fluidity and concentration render it particularly suitable as a food for those winged Insects which lay their eggs singly or a few together on scattered plants of one species, and which must, therefore, spend much time in egg-laying. It is equally advantageous to those which spend much time in building or excavation. Upon the fact that honey can be stored depends the whole domestic economy of Bees and certain Ants.

Honey-sucking is associated with the highest faculties possessed by Insects, and marks, perhaps, the highest phase in their evolution. It is a surprise that Insects with so complex a domestic economy as Wasps and Ants should be able to dispense with it. Like almost all Insects they are fond of honey, but it is seldom their chief food. The Bees have discovered that honey can be converted by chemical change into wax; the gnawing Wasps make paper by chewing vegetable fibres, and use that in their architecture. The Ants have sacrificed their wings, for the sake, it would appear, of carrying on their subterranean work with greater ease. They have paid a heavy price for this advantage, for loss of wings brought about their exclusion from flowers. Ants do get honey, but it is by precarious means and in small quantities. They will drink the sweet excretion of Aphides, if no better supply can be had. Some rifle special honey-glands on the leaves of plants, which appear to have been specially enlarged as a consequence of their visits. Ants are even known to store up honey in subterranean receptacles, the most singular of which are the enormously dilated crops of certain individuals of the community which sacrifice themselves for the general good, and are converted into globular honey-pots.

Some of the honey-sucking Insects which are not Hymenoptera assume so much of the form and colour of Bees or Wasps as to resemble them superficially. Species of Volucella, Eristalis, Syrphus, Bombylius,

Ceria and Conops are often like Bees or Wasps in size, in colour (brown or yellow-banded), in the attitude of the resting wings, in the hairiness of the body, in the narrow waist, and in the telescopic respiratory movements of the abdomen. I have experienced what Réaumur long ago described when capturing some of these Insects. Though the form of the antennæ told me quite unmistakably that I had a harmless Fly before me, I have often hesitated to grasp it, because it looked so like a Bee or a Wasp.

The mimicry of Bees and Wasps by stingless Flies is a proof of the protection furnished by the sting, and of its wide recognition by Birds. If Bees and Wasps were not generally known and dreaded, it would be of no advantage to resemble them. Other flower-haunting Insects may wear the colours of the stinging Hymenoptera, the most striking examples being the Clear-winged Moths, which lose a great part of their wing-scales immediately after emergence, and have the abdomen banded like a Wasp or Hornet. In the same way certain tropical Hemiptera, Beetles and Spiders closely resemble Ants, which are dreaded for the tenacity of their bite, even when unprotected by a sting.

Tennyson as a Naturalist.

Oct. 26, 1895.—This morning I went out early, and found that a touch of night-frost had left its mark upon the shrubs. The lines from *In Memoriam* came into my thoughts:—

> " And Autumn laying here and there
> A fiery finger on the leaves."

Later in the day I rambled through Bolton
Woods. I passed the glowing embers of a fire of
weeds, and stopped to look at the quivering haze.
Again it was Tennyson who had seen the poetical
side of a spectacle so common :—

"All the rich to come
Reels, as the golden Autumn woodland reels
Athwart the smoke of burning weeds."
(*The Princess.*)

In the evening I took down my Tennyson, and
amused myself with noting some of the many passages
which show his knowledge of Nature.

Tennyson is our English Theocritus. It would be
bold to claim that he has excelled the Sicilian idyllist
in charm or knowledge of his art, but it is not ex-
travagant to say that he has given to the grave thoughts
of our reflective age that poetic touch with which
Theocritus was able to brighten the trivial details of a
simple country life. Sometimes Theocritus has been
consciously in the mind of the English poet, as in that
"small, sweet Idyll" of *The Princess.* Perhaps no
English poet since Milton had read to such purpose
in the books of ancient verse, as Tennyson. That
curiosity which led him to glean among old poets has
also made him observant of Nature. It is hard to find
in any other English poet so many of the graphic
touches which show knowledge of Nature and
sympathy with her. Very familiar are the examples
which follow :—

"Those eyes
Darker than darkest pansies, and that hair
More black than ashbuds in the front of March."
(*Gardener's Daughter.*)

"And her hair
In gloss and hue the chestnut, when the shell
Divides threefold to show the fruit within."
(*The Brook.*)

"Bring orchis, bring the foxglove spire,
 The little speedwell's darling blue,
 Deep tulips dashed with fiery dew,
Laburnums, dropping wells of fire."
(*In Memoriam.*)

"I wept, 'tho' I should die, I know
That all about the thorns will blow
In tufts of rosy-tinted snow.
     *   *   *   *
Not less the bee would range her cells,
The furzy prickle fire the dells,
The foxglove cluster dappled bells."
(*The Two Voices.*)

It is not only flowers that Tennyson can use to enrich his verse. I remember one morning after heavy rain climbing the old St. Gothard road as the mists clung to the peaks, and it seemed to me as if one poet only had seen what I then saw.

"The summit's slope
Beyond the furthest flights of hope,
Wrapt in dense cloud from base to cope.

"Sometimes a little corner shines,
 As over rainy mist inclines
 A gleaming crag with belts of pines."
(*The Two Voices.*)

"Leave
The monstrous ledges there to slope and spill
Their thousand wreaths of dangling water-smoke,
That like a broken purpose waste in air."
(*The Princess.*)

Now and then the flash of unfamiliar analogy suggests a thought new to poetry. The lines,

> "Wearing his wisdom lightly, like the fruit
> Which in our winter woodlands looks a flower,"

are the very soul of that Dedication, which, but for the Spindle-tree, would have taken a quite different and less vivid turn. The verses to J. S. reach their highest point when they bring in the long-lasting summer twilight of the northern shores, never turned to such poetic service before.

> "His memory long will live alone
> In all our hearts as mournful light
> That broods above the fallen sun,
> And dwells in heaven half the night."

There is no deep observation but a pleasant humour in the well-known passage :—

> "When the lone hern forgets his melancholy,
> Lets down his other leg, and stretching, dreams
> Of goodly supper in the distant pool."
> (*Gareth and Lynette.*)

Tennyson's natural history allusions have not quite escaped criticism. Mr. J. E. Harting[1] points out two slips. In "The Poet's Song" we used to read :—

> "The swallow stopt as he hunted the bee,"

and in *In Memoriam* these lines occur :—

> "Where now the seamew pipes, or dives
> In yonder greening glade."

The swallow does not hunt bees, and no gull pipes or dives.

[1] *Zoologist*, 1893, p. 145.

That Tennyson's use of natural fact depends upon real sympathy is clear to all who observe how the animals and flowers which throng his written fancies render each its due service. What desolation Tennyson adds to "Aylmer's Field" by the mention of the shy creatures which come back to their ancient haunts, after the wilful lord has undone the last of his race :—

> "And where the two contrived their daughter's good,
> Lies the hawk's cast, the mole has made his run,
> The hedgehog underneath the plantain bores,
> The rabbit fondles his own harmless face,
> The slow-worm creeps, and the thin weasel there
> Follows the mouse, and all is open field."

The stanza,

> "When rosy plumelets tuft the larch,
>   And rarely pipes the mounted thrush,
>   Or underneath the barren bush
> Flits by the sea-blue bird of March,"

sets the larch and the kingfisher of early spring in their corner of the canvas with the sprightliness and the sure touch of Rosa Bonheur. The tapestry which hung the rooms of the Palace of Art is a real gallery of pictures, which many a lover of poetry knows by heart. I quote one verse only.

> "One showed an iron coast and angry waves.
>   You seemed to hear them climb and fall
>   And roar rock-thwarted under bellowing caves,
> Beneath the windy wall."

Among a crowd of other examples which press

for remembrance I find it impossible to pass over
these:—

"So dark a forethought roll'd about his brain,
As on a dull day in an ocean cave
The blind wave feeling round his long sea-hall
In silence."
(*Merlin and Vivien.*)

"O sound to rout the brood of cares,
The sweep of scythe in morning dew,
The gust that round the garden flew,
And tumbled half the mellowing pears!"
(*In Memoriam.*)

"Unwatch'd, the garden bough shall sway,
The tender blossom flutter down,
Unloved, that beech will gather brown,
This maple burn itself away;

"Unloved, the sun-flower, shining fair,
Ray round with flames her disk of seed,
And many a rose-carnation feed
With summer spice the humming air."
(*In Memoriam.*)

"By night we linger'd on the lawn,
For underfoot the herb was dry;
And genial warmth; and o'er the sky
The silvery haze of summer drawn;

"And calm that let the tapers burn
Unwavering: not a cricket chirr'd:
The brook alone far off was heard,
And on the board the fluttering urn:

"And bats went round in fragrant skies,
And wheel'd or lit the filmy shapes
That haunt the dark, with ermine capes
And woolly breasts and beaded eyes;

S

> "While now we sang old songs that peal'd
> From knoll to knoll, where, couch'd at ease,
> The white kine glimmer'd and the trees
> Laid their dark arms about the field."
> (*In Memoriam.*)

Wordsworth, Burns and Shakespeare share this loving appreciation of Nature. I do not find it in Shelley, though the general voice gives it to him. Gray comes near to it once or twice, as here:—

> "The red-breast loves to build and warble there,
> And little footsteps lightly print the ground."

Thomson has his successes, mostly happy words, but they are the gems of a rhetoric whose lustre is not always real. Pope's rhapsody about the moonlight (for Homer has little share in it) won high praise from more than one generation. To us it is nothing but magnificent declamation; no observant person could describe moonlight so.

The examples from Tennyson, which of course illustrate only one side of his poetic endowment, charm us partly by their terse characterisation of what we all know, but never attended to before, but still more by their feeling for the human aspect of Nature. It is not rocks, clouds, flowers and birds which chiefly engage the poet's mind, but the relation of these to the thoughts and hopes of Man; they are intertwined with the history of a man's life. The reality of the observation, the reality of the feeling, save Tennyson from the common faults of those who show knowledge in their poetry; he is never pedantic, nor whimsical, nor cold.

The Structure of a Feather.

Familiar as it is, there are few works of nature which better repay careful study than a feather. Its adaptation to its purpose is complete; it is strong, light, flexible and elastic; its concave surface, which in the case of a quill, is turned towards the inside of the wing, or towards the under-side of the tail, catches the air, while the convex surface allows the air to glide past with little resistance. The feather resembles a host of other natural contrivances in this, that the more we study it, the greater wealth of contrivance we discover. It is wonderful enough when we merely hold it in the hand, and examine it by the unaided eye, but a pocket-lens brings out further and more interesting details, while the utmost refinements are only to be appreciated by those who can command a good microscope and some delicacy of manipulation.

Notwithstanding the utmost diversity in detail, all feathers are constructed upon a common plan. We have feathers with two shafts, feathers with one shaft and feathers with no shaft at all; feathers which bear a stiff and broad vane, feathers which form branching plumes, waving in the gentlest current of air, and feathers which at an early stage of development crumble to powder. Feathers may be used for warmth, for defence, for decoration, for flight. They are of all colours, sizes and shapes. But there is hardly any organ of the Bird's frame more uniform in its early stages of growth.

In the present period of the earth's history, feathers

are absolutely restricted to the class of Birds, and we have no information respecting any extinct feathered animal which was not in essentials a Bird. All known Birds are feathered, just as all known Mammals are hairy.

I can remember something of the excitement which was roused among naturalists by the discovery in 1860 of a fossil feather in the lithographic limestone of Solenhofen in Bavaria. That Birds had existed in the remote Jurassic period was a startling announcement, but how tantalising to have no record of the fact beyond a single feather! The suspense was not long protracted. The very next year the same quarries revealed that fine skeleton of Archæopteryx which is now in the British Museum, and no doubt was entertained that it was this primitive Bird which had yielded the solitary feather found a year earlier.

Take a Bird (a Sparrow is suitable, but any common Bird will do) with all its feathers on, and notice how they are set upon the body. By plucking half the Bird, you can see that the feathers are not placed at equal distances. They are inserted into definite tracts, with bare spaces between. There is a feather-tract along the spine, and a double feather-tract along the front of the body. The sides are to a great extent bare, more in some Birds than in others. If the whole body were closely feathered, the action of the wings would be impeded. But the flightless Ostriches and Penguins are uniformly feathered.

Observe the principal quills used for flight (primaries), and notice that they are inserted into the hand. A Bird's hand is so reduced and mutilated that you will hardly be able to recognise it except by counting the joints of the fore-limb. Nearer to the body comes the long row of secondaries, inserted into the ulna (one of the two bones of the fore-arm). The bases of the quills are bare and separated, to allow freedom of movement during the expansion and folding of the wing, but air is not allowed to rush through the intervening spaces, and so diminish the force of the wing-stroke. The spaces are concealed by the overlapping wing-coverts (upper and under). The tail usually bears twelve quills, and has upper and under tail-coverts. Notice the little "bastard-wing" on the thumb, which perhaps you may not see quite at the first glance. The feet are, in most Birds, bare of feathers and scaly. It is easy to see that feathered shanks and toes would be inconvenient to Birds that run about on wet or muddy ground.

Besides the quills a Bird carries body-feathers of two or three sorts. The larger ones come to the surface, and are hence termed, together with the quills, *contour feathers*. They are compact and glossy, at least in that part which is exposed, and overlap so as to turn the rain. Hidden beneath them are fluffy *down feathers*, which entangle much air. Air is more important than the solid substance of the feathers in preventing the escape of heat. There are also *filoplumes*, feathers reduced to slender, wiry shafts with perhaps a few plumes on one side, or a little tuft on the summit. I suspect that the filoplumes help to prevent the feathers, and especially the down feathers, from becoming entangled one with another. The stiff bristles often scattered through the fur of Quadrupeds are possibly examples of the same expedient. In certain Birds, Herons for example, there are patches of feathers which crumble to powder (*powder-down feathers*). I cannot venture upon any explanation of this curious structure; the powder is often greasy.

Now let us take a single quill, and examine its parts. There is the *barrel*, a hollow cylinder, often transparent, the *shaft*, filled with a white pith, and grooved along its inner side, and the *vane*. The barrel has usually a small hole at its attached end, and a scar upon its inner side, where it joins the shaft. Between these points there can often be seen a chain of dried up husks, often of oval or conical shape. They are easily seen in a goose-quill, especially if one side of the barrel is cut away to expose them. Notice the curvature of the whole quill along its length, and also its more marked curvature from side to side. The concave side is turned towards the inner side of the wing or the under side of the tail; it is usually paler in colour than the other, and marked by the groove along the shaft.

The barrel of a feather is very light, being filled only with air, but it is very strong. I lately took the barrel of a goose-quill, laid it horizontally on supports 2½ inches apart, hung a scale-pan by means of a hook to its middle point and gradually added weights. When the load amounted to 7lbs. the quill began visibly to yield, and at 7½lbs. it collapsed.

I have already attempted to explain the mechanical principle which renders the hollow cylinder so strong in proportion to its weight.[1]

[1] See page 154.

THE STRUCTURE OF A FEATHER 263

In many feathers, especially body feathers, there is a second shaft, the *aftershaft*, which springs close to the scar from the top of the barrel. The aftershaft is usually smaller, often much smaller than the main shaft, but in the Emu and Cassowary it is of nearly the same length. In these large, flightless Birds the feathers serve only for defence and warmth, and here the double shaft is of distinct advantage, allowing twice as many shafts to be crowded into the same surface of skin. I cannot explain why the Ostriches and the little Kiwi of New Zealand have no aftershaft, or none that signifies, but I have long been familiar with *negative exceptions* to every kind of natural contrivance. At first the enquirer is much shaken in his interpretation of a natural structure when he finds that it is wanting altogether in a species which seems to need it as much as any other. But the constant occurrence of such cases where there can be no doubt of the use of the structure leads at length to a settled conviction that Nature has many ways of accomplishing her ends, and can dispense with any organ or any adaptation, often for reasons which are altogether inscrutable to us.

The next thing is to examine the minute structure of the vane. It resembles at first sight a woven fabric. Cut out a square piece, hold it up against the light, and gently pull it across the grain. We see that it is made up of fibres (barbs). The barbs are held together by a multitude of finer fibres (barbules). The barbules will resist a direct pull pretty well, but if the barbs are slid along sideways, they can be detached without violence. They are not truly interwoven, but only hooked together by the barbules.

It now becomes necessary to employ the microscope. Cut out a small piece of the vane, soak it in alcohol to expel the air, then transfer it to glycerine, and tease it out with needles.

We shall then find that the barbs are shaped

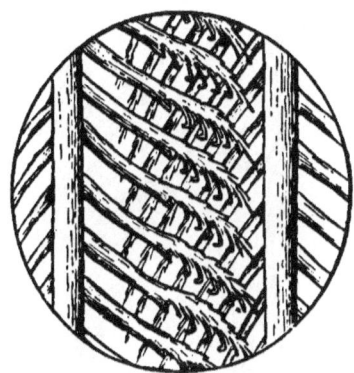

FIG. 63.—Part of a feather, showing two barbs and a number of barbules, slightly separated. The hooks of the distal barbules grasp the proximal barbules of the next barb.

like knife-blades, the back of the blade being turned outwards, away from the body of the bird, and towards the convex side of the quill. Each barb bears a double row of barbules, some hundreds in number. Since the barbs run outwards from the shaft, and the barbules outwards from the barbs, the barbules will be approximately parallel to the shaft. They are only approximately parallel, for they cross one another at a quite appreciable angle. We must now distinguish the two sets of barbules borne upon every barb. There is one set which points towards the base of the quill; these, in accordance with ordinary anatomical nomenclature, may be called the *proximal* barbules. The other set points towards the tip of the quill and will be the *distal* barbules.

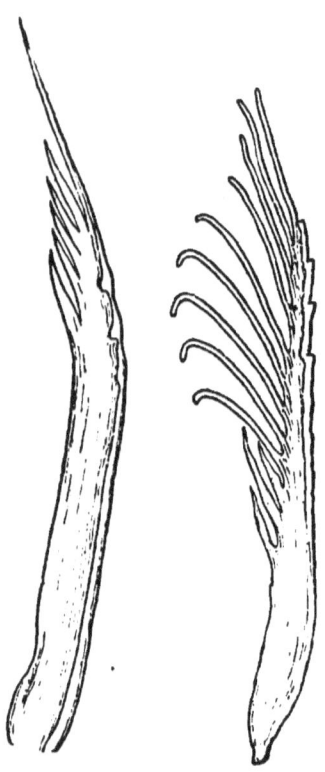

FIG. 64.—Two barbules of a feather. The left-hand one points owards the base of the feather (*proximal barbule*), while the right-hand one points towards the tip (*distal barbule*). The distal barbule bears the hooks.

The distal barbules of every barb overlie the proximal ones of the next barb, crossing several of them obliquely. The proximal barbules have the outer edge turned over at a right angle towards the barb from which they spring, and this projecting edge is, in some feathers at least, scolloped. The distal barbules bear a number of hooks on their inner edge (*inner* here means the side next the body of the bird), and these hooks catch the scolloped edges of several barbules, and hold them strongly, but not rigidly. They can stretch a little and can also slide a little, though the scollops prevent them from sliding too

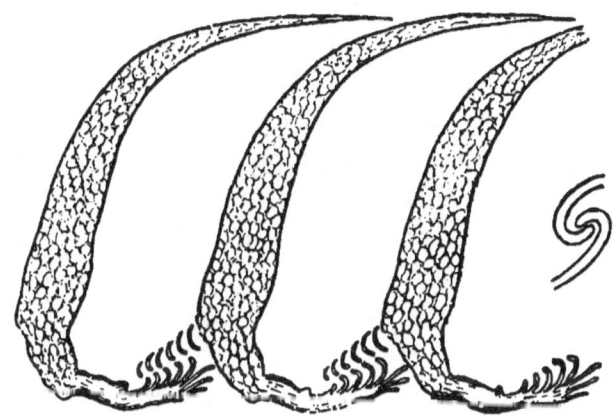

FIG. 65.—Parts of three barbs in section showing their proximal (upper) and distal (lower) barbules. The small diagram illustrates the action of the hooks upon the edges of the proximal barbules.

easily. If they have been gently detached from the proximal barbules, they can be replaced by stroking, and this is no doubt often done when a bird smooths its ruffled feathers with its bill, but rough handling breaks or distorts the hooks, and they never adhere properly again. When the barbules have been studied and drawn, they may be modelled with great advantage. The barbs may be represented by bars of wood, and strips of card cut out to the shape of the barbules may be fixed to saw-cuts made in the sides of the bars. A little trouble bestowed upon the details of the model will not be thrown away; it all helps the perfect understanding of a beautiful and intricate mechanism. In the decorative plumes of many birds the barbules are undeveloped or lose their hooks, and the barbs then become free. A piece of a peacock's feather mounted as a lantern-slide, makes the arrangement of the barbs and barbules quite plain. Double-shafted feathers are easily shown to a number of persons, if mounted in the same way.

A feather cannot be mastered until its development has been studied. Something may be seen of the development of a feather by examination of a moulting Bird, and it is seldom that a bird is not moulting some of its feathers. The new feathers may be seen pushing up through the skin, each enclosed in a thin outer quill, which crumbles gradually away from the tip downwards, and allows the barbs to expand. But the easiest way to get developing feathers is to examine unhatched chicks. Chicks removed from the egg after incubation for nine days and upwards, provide excellent material. But few of my readers, I fear, can command a supply of developing chicks, or know how to investigate them. The work is mainly done by thin sections through the artificially hardened tissues.

There are two layers in the skin of Vertebrate animals, an outer layer (*epidermis*), which is cellular and neither vascular nor sensitive; and an inner layer (*dermis*), which is abundantly supplied with vessels and nerves. Both layers contribute to the formation of the feathers, but the epidermis alone furnishes the formative cells, while the vessels of the dermis bring nutritive substance for the supply of the rapidly multiplying epidermis cells. The first stage in the development of a feather shows a conical elevation of the epidermis, within which the dermis forms a papilla of similar form. As the papilla increases in height, its base becomes sunk to a corresponding extent beneath the general surface of the skin, thus obtaining protection against friction, which would be injurious to a slender column of rapidly growing cells. The epidermic sheath which encloses the dermal papilla increases rapidly in thickness, and the cells arrange themselves in three layers, of which the middle one is much the thickest. After a time the middle layer thins out along one side of the papilla, corresponding to the future inner side of the feather, while it grows in thickness on the opposite side, where the shaft will ultimately appear.[1] At length the line of weakness is broken through, and the upper part of the tube is laid open, forming henceforth a flattish sheet, which is the vane of the feather. Meanwhile unequal deposition of material has given rise to the barbs and barbules, which are due to splitting of what was in a very early stage a continuous conical sheath. The tubular arrangement is retained in the lower part of the feather, which forms the barrel. It is obvious that if a tube is split open along part of its length and laid out as a more or less flattened sheet, while the lower part re-

[1] Where an after-shaft is to be formed two lines of weakness and two thick ridges form.

mains tubular, there must be an orifice where the tube and the sheet join. This orifice exists in every feather, and is called the *umbilicus*; it is usually choked up by a tissue which will be noticed a little later. The apex of the feather is formed first, and may be quite complete while the base is still pulpy; it becomes gradually pushed upwards by the new growth at its base. The outermost epidermic layer forms a cylindrical sheet enclosing the future feather; it adheres strongly to the barrel, but is free from the vane. When the feather first appears above the surface it is enclosed within this outer sheath, from the summit of which a pencil-like bunch of barbs projects. The sheath afterwards dries, and gradually crumbles away from the top downwards, exposing the feathers.

While the growth of the feather is in progress the papilla is relatively large and highly vascular, but shortly before the completion of the barrel, which is the last part to be formed, the papilla begins to shrink. During its retreat from the barrel the papilla leaves behind it successive layers of dry and horny substance, once charged with living protoplasm, and abounding in vessels, but now shrunk to hollow capsules, superposed upon one another. These capsules form a chain, which extends from the base of the barrel to the umbilicus, and in young, unworn feathers may often be seen to project through the umbilical orifice at the base of the vane. At length the feather is completed, and the formative papilla comes to rest. It will however renew its activity periodically during the whole life of the Bird, forming fresh feathers which push out the old ones at the seasons of moult. The papilla also revives whenever a feather is accidentally lost.

Feathers formed in the egg are usually much smaller and simpler in structure than those which are afterwards developed; they form a soft, downy covering in many fledglings.

The colours of feathers are due in part to pigment, but very largely to minute structural peculiarities, such as close-ruled grooves or ridges, which give rise to interference colours, like those of mother-of-pearl or Barton's buttons. That the colours of such iridescent bodies are due to the form of the surface was proved by Brewster, who took casts in black wax, and found that they exhibited the same play of colour. The pigments of feathers are mostly black, brown, red or yellow. Green pigment is extremely uncommon in feathers. The green plumage of a Parrot, if held against the light, or crushed, in some cases if thoroughly wetted with water, turns brown, grey or yellow (Gadow). No blue pigment is known to occur in feathers. White feathers are white because of a multitude of reflecting surfaces, never because of the presence of a white pigment.[1]

THE FALL OF THE LEAF.

Chill October puts an end to the activity of the leaves of our deciduous trees. They cease to be useful

[1] The article on "Feathers" by Dr. Gadow, in Newton's *Dictionary of Birds*, and "The Interlocking of the Barbs of Feathers," by W. P. Pycraft (*Natural Science*, Sept. 1893), may be recommended to those who are able to pursue the subject further.

as food-formers, and it becomes important for the tree to get rid of them quickly and without violence. If the leaves were merely to die in their place, nothing short of a gale of wind would strip the tree, and probably no ordinary gale would suffice, as a fact shortly to be mentioned concerning the Oak and Beech seems to prove. A whole gale, sweeping over a leafy tree, would be attended with loss of twigs as well as of leaves. We see what damage is done by a high wind in summer, when the tree is clothed with firmly adhering leaves. It is much better that the leaf should fall of its own accord in still weather, and return its substance to the soil about the roots, instead of being whirled to a distance. Most of our trees are able to shed their leaves without waiting for them to be torn off, but there are a few unexplained exceptions. The Oak and the Beech keep their leaves long, especially when young. The Turkey Oak keeps its leaves even when it has grown into a large tree. Are these trees adapted to more sheltered situations than others, or are their branches better able to withstand a strain? It is well to put these questions, but I must confess that I cannot answer them.

Leaves about to fall commonly change colour. The chlorophyll either disappears, or is converted into new colouring-matters. The supply of water and sap is arrested, and both leaf and leaf-stalk shrivel. At the base of the leaf-stalk is an enlargement or cushion, and in compound leaves there is often such a cushion to every leaflet. Though the rest of the leaf and leaf-stalk shrink, the cushion remains plump. Let us stop for a moment to consider what is the special use of the cushion to the active leaf.

It is an organ of movement. By means of the cushion the leaf changes its attitude, inclines its surface to catch the light, droops at night, and in some cases droops when touched. The delicate adjustments by which the leaf sets itself in the best position both with respect to light or neighbouring leaves are effected by the cushion. The mechanism of adjustment depends upon the *turgidity* (distention by water) of the cellular cushion. The cells can either absorb water from neighbouring tissues, or give it out, and swelling or contraction follows. There may be swelling on one side and contraction on the other; the swelling may be either temporary or permanent. Swelling on one side causes a leaf-stalk or young stem to incline to the opposite side. Sometimes the cellular tissues of a shoot swell on every side in succession. Then the shoot sweeps round and round in a regular nutation, bowing to every point of the compass in the course of its revolution.

The cushion plays an important part in the fall of the leaf. Here the block takes place, which cuts off the supply of water passing upwards from the stem and roots to the leaf. Here too the *separating layer* forms, which at length severs the leaf from the branch.

The separating layer ends by producing a transverse cut through all the cellular tissues of the leaf-stalk, sparing only the vessels and fibres, though these too it will ultimately break through. The parting of the cellular tissues greatly increases the ease with which the vessels and fibres snap across. Let us suppose that we hold in our hand a slender fishing-rod, and switch it to and fro. It sways in gentle and continuous curves, and unless it is loaded, it will not easily break. Now suppose that we case the rod in a layer of plaster

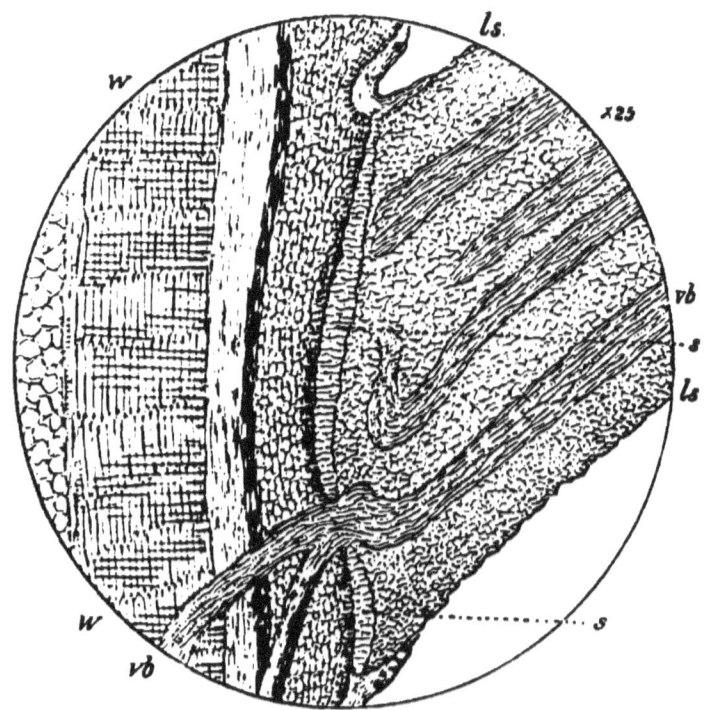

FIG. 66.—Section through leaf-base of Horse Chestnut, before fall of the leaf. ×25. *w*, wood ; *ls* to *ls*, leaf-stalk ; *vb*, vascular bundles ; *s*, corky layer.

of Paris an inch thick, which is ringed, or cut completely through, in one place. What effect will the plaster casing have upon the strength of the rod, and upon its power to resist fracture by bending? Our first thought will probably be that the plaster may

T

increase, but cannot possibly diminish, the strength of the rod. Actual trial, however, proves that this very natural supposition is wrong. The cased rod, ringed in one place, will not stand vigorous and sudden bending, but will snap across at the ring. Rigidity everywhere but in one place is highly unfavourable to that nearly uniform curvature which enables the rod to endure a bending strain without fracture. All the bending is now concentrated upon one place instead of being distributed throughout the whole length of the rod. Mechanical engineers have long recognised that abrupt changes of section greatly increase the tendency of axles and shafts to break across.

The separating layer will therefore weaken the leafstalk, and predispose it to part at one particular place, although it does not pass through the vessels and fibres, which are the principal means of attachment of the leaf. But this is not all. The separating layer contains a mechanism for producing a positive thrust, which comes in aid of the pull of gravity and windpressure, and suffices to part the leaf from the branch, even when it is supported and kept in a perfectly still atmosphere. How this is accomplished I shall try to explain a little later on.

The fall of the leaf is an old subject of enquiry, but the material facts necessary to a satisfactory explanation were not discovered till the year 1859. It happened in that year that the eminent botanist, Hugo von Mohl, spent his long vacation at home instead of at the seaside or in the mountains, and was thus led to observe the fall of the leaf with all the advantages

THE FALL OF THE LEAF 275

of a botanical garden and laboratory appliances. The results of his vacation studies are given in the *Botanische Zeitung* for 1860. Mohl discovered the separating layer by cutting sections through the leaf-stalk when the fall of the leaf was imminent. He found

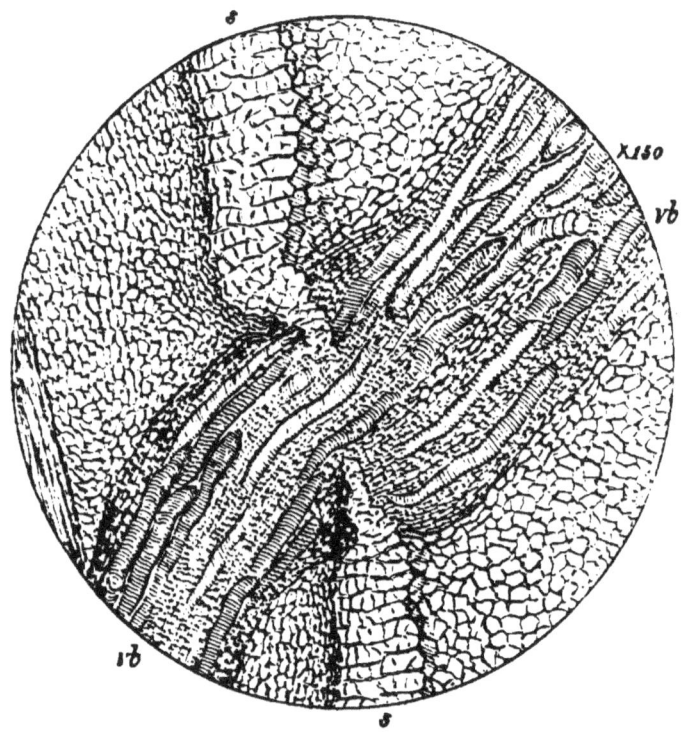

FIG. 67.—Section through leaf-base of Horse Chestnut. × 150 (part of Fig. 66, more highly magnified). *vb*, vascular bundle ; *s*, corky layer.

it to be a thin layer of living and active cells, traversing the cushion. It is charged with abundant living protoplasm, contains many starch grains, and appears in the midst of cells which are almost empty and well-nigh dead. The separating layer can often be picked

T 2

out in sections by the naked eye, especially if iodine solution, which gives a characteristic blue colour to starch, is applied. It forms gradually, extending across the leaf-stalk from without inwards.

The separating layer consists of growing tissue, which absorbs whatever nourishment it can draw from the neighbouring cells, and displays a short-lived activity. New cell-walls, parallel to the direction of the layer, soon appear, and in a few days or even in a few hours after its first appearance in a recognisable form it becomes divided into three tiers of cells. Each cell is very thin or low, in proportion to its length and breadth. When the right moment comes, the middle tier of cells breaks down, the cell-walls being most likely converted into a kind of thin mucilage by a change of which many other examples could be furnished, and thus the cellular tissues of the leaf-stalk are severed.

Since Mohl's discovery further investigation has brought to light not a few interesting details. Van Tieghem and Guignard[1] have pointed out the significance of certain peculiar brown cells, previously seen by Mohl, which stretch across the leaf-stalk near the separating layer. They are cells which have become lined with cork, and thus rendered impervious to water and watery fluids. Before the leaf is shed the cellular tissues have their supply of water and sap cut off. The vessels, however, still remain open, for they are wanted to discharge whatever useful fluids the worn-out leaf may still contain. After the leaf has fallen, the vessels too may be sealed by corky sub-

[1] *Soc. Bot. de France*, tom. 29 (1882).

stance. The corky layer is often formed months before the fall of the leaf. It is usually a little lower down than the separating layer.

Van Tieghem and Guignard also tell us, though I think that they did not first discover the fact, that the changes in the leaf which precede its fall may be artificially induced any time after Midsummer. It is only necessary to cut a branch, and keep it in a still, moist atmosphere. Shutting it up in a botanical collecting-box is a very convenient method, which will cause the complete formation of a separating layer in a week or less. It is not, therefore, necessary to stay at home during the long vacation in order to study the phenomena of defoliation with all the conveniences of our own laboratory or study.

Van Tieghem and Guignard observe that when the middle tier of cells in the separating layer deliquesces, the exposed cells of the neighbouring tiers begin to bulge. This points to their turgid condition. Our authors believe that increasing turgidity at length causes the two tiers to press against each other with sufficient force to rupture the vessels and fibres. Thus the last attachment of the leaf is severed, without shock and it may be in perfectly still air, and the leaf falls gently to the ground.

Mohl observed that during a frost in early winter many leaves fell though the air was perfectly still. On close examination he found that the sap in the separating layer had frozen to a thin plate of ice, which forced the tissues apart, just as ice in the crevices breaks up the clods. When the ice melted after sun-rise the leaves fell at once.

Autumn Winds and Winter Floods.

Nov. 23.—For weeks past there has been a succession of south-westerly gales with torrents of rain. The withered leaves have been whirled away, the roads are deep in mire, the river is in flood. All through the autumn the grass has been growing, and flower buds have been opening months before or after their usual time.

The ocean of air which rests upon the earth is a most unstable thing, sensitive to the slightest change of temperature. Inequalities of temperature create movement, and the movement once set up does not easily subside. The currents of the air, like currents of water in a deep pool, seldom take a straight course, but circle, or boil up from the depths and then plunge down again. The eddies of the air, like those of the river, have a tendency to keep to certain tracks. Land and sea are fixed things, and these determine to some extent the distribution of the temperature and the set of the winds. For weeks together cyclones go whirling along from S.W. to N.E. between Spain and Iceland, nearly always passing to the north of our islands, but swerving a little from time to time. They bring with them the moisture of the ocean-air, and something of the warmth of lower latitudes.

A week ago the rivers rose to an unusual height. The banks were swept bare in many places. Trees were uprooted, and felled trees set afloat. Standing on a bridge to watch the rushing torrent I could see trunk after trunk shoot past. The stream has now fallen again, and I have been to watch the effects of

AUTUMN WINDS AND WINTER FLOODS 279

the flood. On one low flat near the river I found great patches of refuse, broken twigs, cases of caddis worms, here and there a chrysalis or a cocoon, and abundance of seeds. Among many unknown fruits and seeds I could see a great many Alder-nuts, which happen to be familiar to me.

Alder grows by preference on the banks of streams, and during the gales of autumn and winter the ripe nuts are shaken out of the cones. Many of them must fall from the overhanging boughs into the water, and be swept down the stream. Do they sink or swim in water? It was a simple thing to try. I threw a number of the nuts into a beaker of water, and found that they all floated. They went on floating all through the winter, and many of them germinated on the surface in spring or earlier. Do all seeds swim in water? I went to a seedsman and bought a dozen packets of flower-seeds, taking the first which came to hand without selection. All sank in water except a small proportion of each sort (probably dead seeds) which contained air and floated. Among the rest were the so-called seeds of two Composites, which floated. But these were not mere seeds. They were fruits invested by the withered husks, which enclosed plenty of air.

On cutting open an Alder-nut the wall was found to be excavated by numerous small cavities. I wished to ascertain whether these were air-tight compartments or not. I therefore exhausted the air from a receiver which contained some entire and some broken Alder-nuts floating on water. None of them sank, though the air was kept exhausted for a long time. I concluded that the cavities were separate and airtight.

The long flotation of Alder-nuts on water suggested that they contained some resinous or other water-repelling substance. Dr. J. B. Cohen, of the Yorkshire College, was good enough to examine them for me. He says: — "About $4\frac{1}{2}$ grams of Alder seeds were dried in a steam-oven and extracted with ether. They lost in the first process 17·1 per cent. of water, and in the second 2·4 per cent. of extractive matter. The substance extracted formed a perfectly solid and brittle, light-yellow mass. It gave none of the reactions for resin, but on heating melted and evolved a smell closely resembling that of hot linseed soil. The seeds freed from oil floated on water, but some of them sank after a time. The majority, however, floated after four days' soaking. On the other hand, when the seeds freed from oil were cut in halves, and soaked in water, and the air then extracted by exhaust-

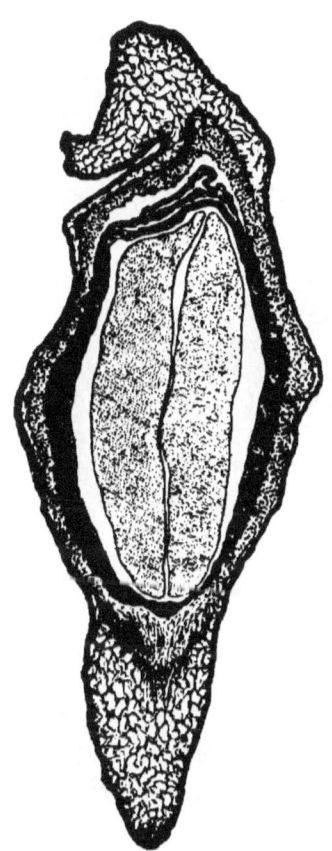

FIG. 68.—Section through Alder-nut, showing the seed in the centre and the porous husk outside. Magnified.

ing the vessel under the air-pump, all the seeds without exception sank after twenty-four hours." The long-continued flotation of Alder-nuts is therefore to be attributed to the numerous air-tight compartments of the wall or shell, and to the oily matter which renders them incapable of wetting. We shall shortly see that both precautions are employed in the case of another fruit which is dispersed by water.

I next turned to Dr. H. B. Guppy's paper on the

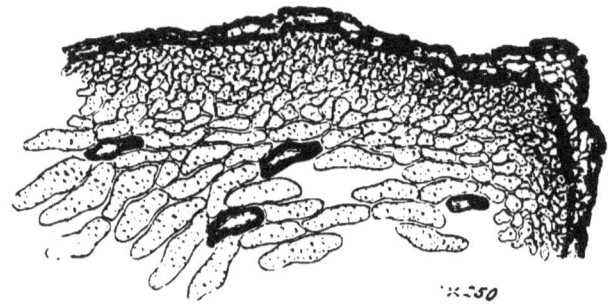

FIG. 69.—Part of the porous husk of an Alder-nut, highly magnified. The dark spaces are filled with air.

River Thames as an agent in plant-dispersal,[1] which contains many curious facts. He tells us, as the result of his long-continued and laborious enquiries, that not only in autumn but in winter and spring the rivers carry down much vegetable drift. It is not usually swept down at once to the sea. Winds blowing across the river set up a surface-flow, by which the drift is often lodged among the reeds or embayed in sheltered creeks. Floods throw the drift upon the banks, where it may rest for weeks and months until another flood picks it up. Eddies detain

[1] *Journ. Linn. Soc., Botany*, Vol. XXIX., p. 333 (1893).

it, perhaps for many days together, in the same place. In the lower reaches of the river the drift comes under the influence of the tides, which carry it to and fro for a long time. During this protracted flotation, seeds, seed-vessels, and even broken-off fragments of living plants may retain their power of germination or renewed growth. Some, like the nuts of the Alder, float a long time and germinate in spring at the surface of the water. The seedlings of such plants would readily establish themselves whenever they happened to be stranded in a suitable place. The seeds and seed-vessels, which float for months in the river-drift, nearly always float equally well in sea-water, and afterwards germinate, as Dr. Guppy ascertained by actual experiment. Ice sends great numbers to the bottom after the thaw, but many are not injured even by repeated freezing. Some seedlings even gain fresh vigour in the ice, and will put forth their leaves during the daily thaw, though every night they are frozen up again.

The floating drift by no means includes the seeds of all the common water-plants of the river. It would be nearer the truth to say that it includes none of them. Water-lilies, the Water Persicaria (*Polygonum amphibium*), the Water Ranunculus, the Water forget-me-not and other plants which live actually in the water are unrepresented in the drift. Their seeds have little or no floating-power. These plants probably owe their dispersal to birds. Charles Darwin tells us that hard seeds pass uninjured through even the digestive organs of a Turkey. He picked up in his garden twelve kinds of seeds from the droppings of small birds, and some of these, which were tried, germinated. The castings of Hawks and Owls often contain seeds capable of germination. Fishes eat the seeds of many land and water plants, and are themselves often eaten by birds. When seeds were stuffed into the stomachs of dead fishes, which were afterwards given to Fishing-eagles, Storks and Pelicans, the seeds were afterwards thrown up or passed out, and several of them were able to germinate. Even large insects, such as Locusts, transport living seeds in their intestines.[1]

The winds, which bring the rain and swell the rivers, are another means of dispersing seeds. The plumed fruits or seeds of the Thistle, Dandelion, Willow and Bullrush, the winged fruits of the Elm, Ash, Sycamore, Lime, Birch and many Umbellifers are carried over the fields by high winds, and those which are small and light may be carried very far indeed. I have seen plumed seeds settle down on the waves at a distance of more than twenty miles from shore, and if my opportunities of observation had been better I could no doubt give much more striking cases.

Less frequent modes of dispersal are entanglement in feathers, fur or wool (to facilitate which many low plants have their fruits or seeds hooked), and mechanical ejection, such as is practised by the Violets, Geraniums, Furze and many others. Here the distance to which the seeds can be directly conveyed is very limited, often only a few feet, but dense crowding at least is avoided.

[1] *Origin of Species*, ch. xi. Many other facts of the same order are given in the original.

The winds which blow steadily over great expanses of sea set up currents, which often carry floating objects to great distances. Among these are many well-known drift-fruits, such as the Coco-de-mer (the Lodoicea of the Seychelles), the Sea-apples or Sea-cocoa-nuts of the West Indies (fruits of the Bussu palm of Trinidad and Brazil), and the Sea-beans (*Entada scandens*) which are cast ashore in all parts of the world. Linnæus long ago noted that tropical fruits and seeds, in some cases capable of germination, were frequently stranded on the coast of Norway. One drift-fruit, often cast ashore in the West Indies and elsewhere, is particularly interesting, first because it exhibits the same structural peculiarities which fit the Alder nuts for dispersal by water, and secondly because, though it is often cast up on the sea-shore, its native country and the tree which yielded it were only discovered after three centuries of inquiry. In *Nature* for Nov. 21, 1895, I find an article entitled "A Jamaica Drift-fruit," in which Mr. D. Morris, Assistant-director of the Kew Gardens, clears up this ancient mystery.

The fruit in question was first described and figured by Clusius in 1605. After that it was repeatedly discovered as a waif upon tropical shores, and once (in 1887) in Bigborough Bay in the south of England. From the large collections preserved at Kew, Mr. Hillier was at length enabled to infer that the fruit was probably referable to the small order of Humiriaceæ, which contains trees or shrubs mostly with balsamic juice, entirely confined to tropical America, so far as was then known. This led to further inquiries, and at length to the recognition of the plant by Mr. Hart, Superintendent of the Botanic Garden at Trinidad. There the tree still grows. It

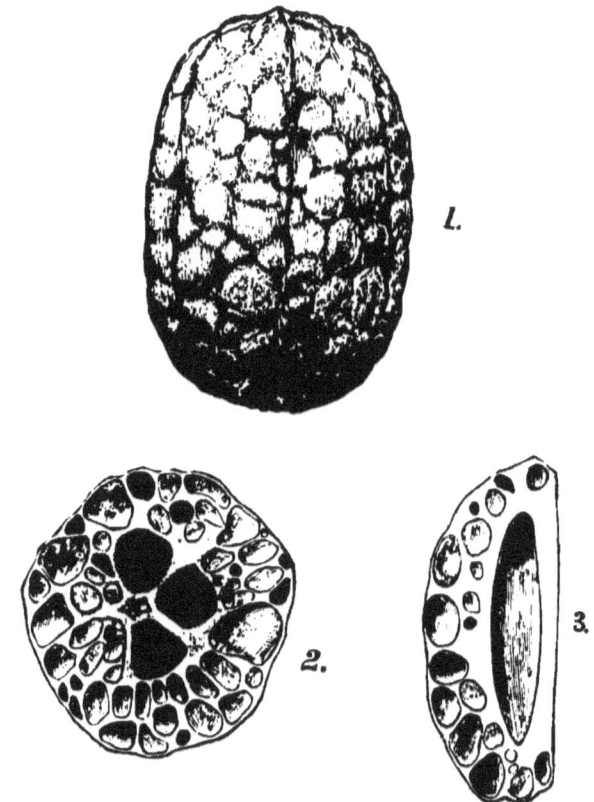

FIG. 70.—Fruit of *Saccoglottis amazonica*. From paper on a Jamaica Drift-fruit by D. Morris (*Nature*, Nov. 21, 1895). 1, fruit with fleshy exocarp removed, as in drift-fruits. 2, cross-section, showing the numerous cavities. 3, longitudinal section.

was botanically described by the former Director of the Trinidad Botanic Garden, Dr. Crüger, in 1861. The tree is named *Sacoglottis amazonica*; it is rare in

Trinidad, but more abundant in the delta of the Amazon.

It is highly satisfactory to be able to trace a long-known drift-fruit to its native home. The interest is increased by the peculiar structure of the fruit. It is covered externally by a thin fleshy envelope, within which is a shell, excavated by numerous large and

FIG. 71.—Fruit of *Sacoglottis amazonica*. From *Nature*, Nov. 21, 1895, after Crüger.

irregular cysts, which contain air and some resin. Hence the fruit is very buoyant, easily impelled by wind, and not easily water-logged. The tree grows near running water, which can transport the fruits to the sea, and so to distant shores.

The Sacoglottis-fruit, in spite of its protection from sinking, decay or destruction by animals, is not known to have established itself in a new area within recent times. Its transport from Brazil to Trinidad may have been effected by flotation, and very likely was, but proof of the fact is inaccessible. This reminds us that many things are necessary for establishment in a new area besides mere transport of the living plant or animal. Innumerable plants and a considerable number of animals reach our shores every year from distant parts of the world, some borne by currents, some by the wind, many more by man himself. Of these we may shortly say that none survive when fresh supplies are cut off. It is just possible indeed to point to the Anacharis of our ponds and streams as a casual invader which holds its ground here, but I know of no second instance. Our experience is however exceptional. The British Islands are crowded with dominant species, and there is no room here for chance immigrants. It would be very different with the weeds or common animals of a wide continent which chanced to be cast ashore upon a long-isolated island. There, the material difficulty of transport once overcome, the invaders would have a fair chance of survival. Keeling Island and many other examples show that wide stretches of sea may be crossed, and that numerous migrants may at length establish themselves permanently in a new soil. The population of the earth, both animal and vegetable, would be very different from what we now see, if it were not for the means of dispersal provided by the winds and waves, and for that adaptability to external conditions which enables plants and animals to employ those means for their own purposes.

The Shortest Day of the Year.

Dec. 21.—The sun rises in London at 8 h. 6 m., and sets at 3 h. 51 m. To-day there is less than eight hours of full daylight, more than sixteen hours of night and twilight. Even at noon the sun has little power. His rays strike us at a low angle, $15°$ only. On the longest day he attains a height of $62°$, more than four times as high as at noon on the shortest day.

The sunshine is not only brief but faint, because of the small elevation of the sun at noon. Take any definite part of the earth's surface, such as a particular field. If that were turned full towards the sun it would receive the greatest possible number of rays; if turned edgewise, it would receive none; for any intermediate position it would receive more or fewer according to its inclination, the amount being proportional to the sine of the angle. At $62°$ it would receive ·883, at $15°$ only ·259 of the full amount, which is taken as unity.

When the sun is low, his rays pass very obliquely through the earth's atmosphere, and much light and heat are absorbed. If the depth of the atmosphere is taken at unity, a ray passing through it at $62°$ from the horizontal plane will be 1·133 long, 3·864 long at $15°$.

The shortest day is not on an average of years the coldest day, nor the longest day the hottest. We have to take into account the effect of foregoing temperature. In summer the surface of the earth warms up steadily day by day, and the maximum falls a little later than the longest day, viz. about July 14—16. In winter the surface of the earth cools a little every day, and the minimum falls about January 8—11.

The naturalist is little abroad in December. It is Nature's long vacation, and many works of the sun have perished or are to all appearance dead. It is the time to enjoy the works of man. The new books lie in the shops; the fireside and the study-lamp shine bright "now that the fields are dank and ways are mire." For those who care nothing about books there are the theatre and concert and ball. Even the street-lamps and the roll of carriages help to dispel gloom.

No doubt there is much to be seen and studied after the leaves have fallen and before the blood stirs again in the veins of Nature. The threads of the web of Life are being gathered up. Careful packing and housing there must needs be; there are seeds to be protected against frost, pupæ to be hidden where the birds cannot find them; the soil has to be fertilised against a fresh crop. In December, dark, cold and wet, multitudes of living things hold their life somewhat as did the shipwrecked Ulysses, heaped over with leaves on the Phæacian shore, "as when a man has hidden away a firebrand among black ashes on a lonely farm, where there are no neighbours, carefully saving the seed of the fire, that he may not have to go in search of the kindling spark."[1] Life is still warm in the branches that seem so dead, in the fallen

[1] *Odyssey*, end of Book V.

fir-cones, in the invisible eggs of Insects, in countless particles that we cannot distinguish from dust and sand and rotting leaves. The very ground, frozen hard and covered with grey stubble, is rich in hope, and holds the promise of many a spring to come.

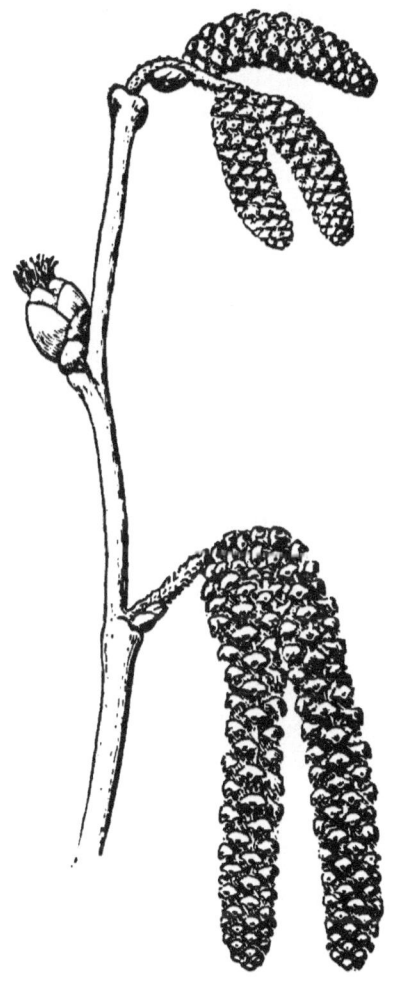

FIG. 72.—Flowering branch of Hazel (see p. 81).

INDEX

INDEX

ADDER'S tongue, 1
Alder, flowers of, 73, 76
 nuts of, 79, 279
Alpine and maritime plants, 225
Ancients, their sentiment for scenery, 230
Angora, 39
Animals, cold endured by, 7
 recovering after freezing, 6
Ants, 251
Archæopteryx, feather of, 260
Autumn Winds and Winter Floods, 278

BEAMSLEY Fell, Corn-rigs of, 103
Beech, bud of, 126
Bilberry, 227
Birch, flowers of, 80
 fruits of, 74, 86
Birds, fruit-eating, in frost of 1895, 31
 in mid-winter, 26
 in snowy weather, 27
 perishing of cold, 27
 which sing in winter, 26
Blooms, Midsummer, 142
Bone, Mechanics of, 155
Botany of a Railway-station, 137
 language of, 208
Bryony, 236
Buds, 121
Bulb of Onion, 72
Bulbs, 65
Buried in the Snow, 25
Butterflies, Cabbage White, 158
 Hair Streak, 179

 Peacock, 178
 Small tortoise-shell, 178
 Swallow-tail, 180

CABBAGES and Turnips, 183
Cabbage White Butterflies, 158
Caddis-worms in Wharfedale, 36
Cat, 38
Caterpillar on Snow, 1
Cold endured by animals and plants, 6
Combs of Animals, 49
Corm of Crocus, 66
Corn-rigs of Beamsley Fell, 103
Cow and Calf, Ilkley, 32
Cranberry, 217
Craven, Agriculture in, 105
Crocuses, 64
Crowberry, 211
Cuckoo, 107
Cup and Ring Marks, 34

DEPTH to which the ground freezes, 28
Desert plants, 224
Dog, 38
Duckweed, 192
Dytiscus swimming beneath Ice, 3

EVERGREEN Plants, 226

FALL of the Leaf, 270
Feather, Structure of, 259
"February fill-dyke," 47
Flower-haunting Insects, 248

INDEX

Flowering Currant, bud-scale of, 123
Frost, depth to which it reaches, 28
   effect on injurious Insects, 32
   effect on marine animals, 31
   fruit-eating Birds in, 9
   Furze in, 31
   of 1895, 30
   water-mains in, 29
Fruits and seeds dispersed by currents, 284
Fungi, habitats of, 139
Furze in frost of 1895, 31

GIRDLED pupæ, 179
Gossamer, 240
Grasses, 145
Gray's Tour in north of England, 231
Grouse in winter, 27

HABITATS of Fungi, 139
   of Insects, 139
Hair-streak Butterfly, 179
Harvey on Insects, 165
Hay-time, 143
Hazel, flowers of, 73, 80
   fruit of, 81, 87
Heather, 211
Hoar-frost, 15, 21
Honey-sucking Insects, 249

ICHNEUMONS, 182
Insect-pupæ in winter, 1
Insects, flower-haunting, 248
   habitats of, 139
   transformations of, 164
   in mid-winter, 3

JENNER, 108

LAGRANGE'S maxims, 199
Landslips, 33
Lapland, cold winds of, 223
Leaf, Fall of the, 272
Leaves of moorland plants, 210
Leprosy, 191
Lilac, bud of, 128
Ling, 211

MACALISTER on mechanics of Bone, 155
Malfrighi on Insect-transformations, 166
Maritime and Saline Plants, 225
Mayerne, 165
Meloe, 89
Microgaster, 183
Midsummer Blooms, 142
Mimicry of Bees and Wasps, 252
Moon, 54
Moonwort, 2
Moorland Plants, 208
Moufet's Theatrum Insectorum, 165
Mountains, the Love of, 229
Muscatel, flower-bud of, 135

NARDUS, 215
Negative exceptions, 79, 229

OIL-BEETLE (Meloe), 89

PALÆARCTIC plants and animals, 206
Paramos, vegetation of, 219
Peacock Butterfly, 198
Petrarch, his ascent of Mont Ventoux, 230
Plants, cold endured by, 6

RAILWAY-STATION, botany of, 137
Réaumur on pupæ of Butterflies, 177
Redbreasts, 26
Reversed Spiral, 236
Rock-crystal of Alps, 24
Rousseau on Swiss scenery, 231
Routine, 199
Ruminant colon, 239
Rush, 218

SACCOGLOTTIS AMAZONICA, 284
Saddleback, ascent of, 232
Scurvy, 191
Shortest Day of the Year, 288
Simon's Seat, 229
Simulium-larvæ in winter, 3
Sitaris, 101, 102

INDEX

Snow, air and water in, 24
  buried in the, 25
  flakes, 11
Stipules, 127
Summer twilight, 140
Swammerdam on Insect-transformations, 166
Sycamore, 37
  buds of, 121
  leaf of, 126

TADPOLE, intestine of, 239
Tennyson as a naturalist, 252
Thecla, 179
Tortoise-shell Butterfly, 178
Transformations of Insects, 165
Turnips, 183
Twilight, Summer, 140

UNDER the Crags, 32
Unisexual flowers, 83, 88

WATER-MAINS in frost, 29
Weeds, 201
Wettest months, 47
White, Gilbert, 3, 8
Willow, flowers of, 82
  fruits of, 84
Wind, its effect upon leaves, 220
  and the dispersal of seeds, 283

XENOPHON on Hunting, 230
Xerophilous plants, 224

YORKSHIRE FOG, 145

www.ingramcontent.com/pod-product-compliance
Lightning Source LLC
Chambersburg PA
CBHW021957220426
43663CB00007B/854